INDICATOR PLANTS

USING PLANTS TO EVALUATE THE ENVIRONMENT

PAUL L SMITH

In memory of my parents
C.H.H. and E.L. Smith, a craftsman and a cook

Cover photograph courtesy of Carol E Jenner

> ***"I can see nothing," said I, handing it back to my friend.***
>
> ***"On the contrary, Watson, you can see everything. You fail, however, to reason from what you see. You are too timid in drawing your inferences."***

Extract from a conversation between Holmes and Watson from
'The Adventure of the Blue Carbuncle'
– Arthur Conan Doyle,Strand Magazine 1892[1].

Plants and their assemblages do not occur randomly in the environment. Their distribution is dictated by overarching environmental variables, which in many cases are well known. By knowing the species and something of its tolerances, the nature of the environment in which it is found can be deduced. Particularly useful examples are known as 'indicator plants'; these are keys to a broader understanding of the ecosystem and form the subject of this practical guide.

[1]This text from the works of Arthur Conan Doyle has been reproduced in the understanding that the work is no longer under copyright in the UK.

Contents

1. INTRODUCTION

This book was written with three sets of people in mind:

- Junior ecological consultants who want to improve their Phase 1 Habitat Surveys[2] with botanical target notes useful in ecological assessment.
- Students of ecology who need to use plants to interpret the landscape around them and understand habitats in greater detail.
- Amateur naturalists and walkers who seek to enrich their experience of the countryside through a knowledge of plants.

Compiled from a range of published sources and from the author's personal experience, it is intended for use primarily in the lowland English countryside. The book takes the form of two tables and an exercise. Relevant published sources are cited in the tables and fully referenced at the end of the document.

The guide looks at how plants may be used to 'read the environment', essentially treating them as indicators of the state of the habitat where they are found. In this respect it is important to note that the use of plants in such a way is based upon a measure of probability. For example, some plants are mostly found growing on chalk but may occasionally be found on what appear to be acidic substrates due to the influence of a calcareous run-off. Ecotypic variation in the environmental responses of a species can also result in unexpected distribution patterns. The rule is that the presence of certain plants in an area may indicate that certain environmental conditions are prevailing there but they do not necessarily prove that they are. They do however provide clues to environmental conditions and with this perspective they are of considerable use to the environmental investigator. This book is therefore presented as a *vade mecum*[3], and is intended as a handy little text for ecologists and naturalists active in the United Kingdom.

HISTORY OF THE IDEA

The use of plants to 'read the landscape' is not a new idea. 'Indicator plants' have been used in many types of investigation. Mineralogists (geobotanists) have relied on the presence of certain plant species to locate metalliferous deposits in the soil. For example, several 'copper flowers' have been identified and have been used for mineral exploration. The plant *Becium homblei* was used in the 1950s to indicate copper bearing areas in Africa (refer to Baker & Brooks, 1988 for an account).

Ecologists regularly take note of plant assemblages to indicate the pH of the surrounding soilscape. The presence of heather (*Calluna vulgaris*) indicates an acid soil of low base status; the presence of yellow-wort (*Blackstonia perfoliata*) indicates the opposite. The presence of certain plant species such as herb-paris (*Paris quadrifolia*) and toothwort (*Lathraea squamaria*) in a woodland have come to be regarded as an indication of the site's ancientness (e.g. Hornby & Rose, 1986).

---

[2]*sensu* Joint Nature Conservation Committee (JNCC, 1993). The Phase 1 habitat classification and methodology as published by the Nature Conservancy Council in 1990 (reprinted by JNCC in 1993, 2003) continues to be used as the standard 'phase 1' technique for habitat survey across the UK.

[3]A handbook or other aid carried on the person for immediate use when needed (C17: from Latin, literally: go with me). Collins English Dictionary. 3rd Edition,1994. HarperCollinsPublishers.

By the nature of their ecology and responses to environmental variables, *plants and their assemblages are not scattered randomly in the environment*. Their distribution patterns are a direct response to overarching environmental parameters, which limit the processes of dispersal and natural succession. In many cases, for individual plants these limiting factors are well known and have been described in forms such as Ellenberg's indicator values (refer to Hill *et al.*, 1999, 2004). This being the case, the underlying 'code' may be reversed and experienced field botanists have used this almost subconsciously as a means of reading the botanical indicators around them. This often leads to nebulous but meaningful statements such as; '*it is a good habitat*', '*that area of grassland was very interesting*', '*it is an old hedgerow*' and perhaps most frustratingly unprofessional of all '*that site is special*'[4]. This guide draws together some of the signals ecologists are responding to when these statements are made and presents them in tabular form. It is a prerequisite of a good indicator plant that it is easy to identify in the field. The present account is also largely restricted to relatively common examples, which the author has personally found useful, with a view to making the document of value in interpreting a broad range of frequently encountered habitats.

Please note that these indicators have been compiled with reference to many published sources and these have been noted within the table. The present author has interpreted and summarized such information with reference to personal field experience and must accept responsibility for any errors arising.

It is important to remember that research and critical observation can overturn traditionally held views. For instance, the fern *Ophioglossum vulgatum* has long been considered an indicator of ancient, undisturbed grasslands. Recent work has shown however that it can appear in disturbed roadside environments (refer to Crowther *et al.*, 2009). The botanical investigator should keep an open mind when deducing meaning from the presence of plant species in semi-natural or anthropogenic habitats. Literature is sometimes shaped by intellectual fashion and the reader is encouraged to follow the references cited with a view to understanding any challenges to accepted wisdom which this book presents.

[4]To refer to a habitat as *special* is about as ecologically meaningful as a physician informing a sick patient, after a rigorous examination, that they are '*a bit unwell*'.

2. THE INDICATORS

The indicators have been set out in two tables. Table 1 includes a range of species, arranged in alphabetical order by their scientific names and provides a synopsis for each, distilling what their presence indicates about the locality where they are found. It also includes, where appropriate or relevant, an indication of the broad Phase 1 Habitat (JNCC, 1993) where the species may be expected, an indication of corresponding National Vegetation Classification (Rodwell,1991a,b, 1992, 1995, 2000) communities in which it has been recorded, the published Ellenberg values for the taxon (usually Hill *et al.*, 1999, 2004, 2007) and reference to its strategy under Grime's (1979) Competitor-Stress tolerator-Ruderal (C-S-R) theory (*cf.* Grime *et al.*, 1988, 1990, 2007). Table 2 includes a range of habitats and environmental features commonly encountered in the lowland English countryside and provides indicators of use in their investigation.

It is essential to note while using this book that these indications are based on probabilities only. They have a scientific basis but are not scientifically rigorous in themselves and contribute to habitat assessments what ornithologists refer to, when identifying small brown birds from a long distance, as 'jizz'[5]. This guide seeks to provide the beginner with what might be termed 'habitat jizz'. There will be many exceptions to each rule and any deductions made or conclusions drawn should be substantiated with additional evidence.

NOTES ON ELLENBERG VALUES AND GRIME'S CATEGORIES

Any account of indicator plants should make reference to existing systems and sources of relevant information. Two sources drawn heavily upon by the present guide are the works of Hill *et al.* (1999, 2004, 2007) dealing with Ellenberg values and derived indicator values for British plants and Grime (1979) and Grime *et al.* (2007) dealing with the ecological strategies of plants. For convenience, these systems are briefly introduced below but the reader is referred to the original sources for a more detailed understanding.

Ellenberg defined a set of indicator values for the vascular plants of central Europe (Ellenberg, 1979, 1988; Ellenberg *et al.,* 1991). These have been widely used in central and parts of western Europe and have been modified for use in Britain by Hill *et al.* (1999). Hill *et al.* explain that the basis of indicator values is the realized ecological niche, i.e. the fact that plants have a certain range of tolerance of temperature, light, soil pH, etc. For this reason, the local flora can provide much useful information regarding the ecological conditions pertaining at a site. In the introduction to *'Ellenberg's Indicator Values for British Plants',* Hill *et al.* (1999) consider that the flora may indicate quite a narrow range of environmental conditions. For example, the presence of rhododendron (*Rhododendron ponticum*) indicates an acid soil whereas the presence of small scabious (*Scabiosa columbaria*) indicates a basic or alkaline substrate. Indicator values seek to encapsulate this information. For example on the scale R (soil reaction), *R. ponticum* has the value 3 and *S. columbaria* has the value 8. These values are not mean pH values but are on an arbitrary scale reflecting soil pH though not directly based on measurements.

Five Ellenberg values are provided for flowering plants (Hill *et al.*, 2004): light (L), moisture (F), soil reaction (R), nitrogen (N) and salt tolerance (S), with an additional value for heavy metal tolerance (HM) for bryophytes (Hill *et al.*, 2007). Summary tables for the meaning of these indicator values are provided in Appendix 1.

The indicator values provided by Hill *et al.* (1999, 2004, 2007) were based on those of Ellenberg *et al.* (1991) and derived by a variety of means (refer to Hill *et al.*, 1999 for a detailed account). They are reported as being a mixture of objective results based on calculation and subjectively derived values based on field experience and published sources. Values for bryophytes were derived from statistical analysis of quadrat data and

[5] The Oxford English Dictionary 2nd Edition (1989) online Version (November, 2010) www.oed.com, defines 'jizz' as 'the characteristic impression given by an animal or plant'. It is mimicked by the multivariate statistic procedure of Principal Components Analysis whereby many variables are distilled into recognizable clusters.

associations with vascular plants, information from published and unpublished sources and personal experience (refer to Hill *et al.*, 2007).

A measure of discretion should be used when applying any type of indicator approach with living organisms because biological entities are complex and may respond to the environment in different ways in different places. The use of indicator plants in environmental interpretation has been shown to be a valid technique by empirical use over many decades, but it is not definitive in the sense that chemical titration may be seen to be definitive. It is only indicative.

The work of Grime *et al.* (1988, 2007) is explained in *Comparative Plant Ecology: A Functional Approach to Common British Species*. An abridged version of the first, 1988 edition to this work appeared in 1990 (Grime *et al.* 1990) and provides a simple introduction to the subject. Essentially, Grime *et al.* recognise plant attributes which, by becoming the foci of conflicting selection pressures, limit the distribution of genotypes, populations and species. The authors consider crucial to this approach the hypothesis that ecological specializations important to the present or past success of an organism may frequently involve the assumption of genetic characteristics which render the plant unsuited to life in 'other' environments, hence restricting its current ecological range (Grime *et al.*, 1990).

Grime *et al.* (2007) assign ecological strategies, or functional types, to plant species, i.e. recurrent types of specialization associated with particular habitat conditions or niches. Grime (1979) defines a strategy as 'a grouping of similar analogous genetic characteristics which recurs widely among species or populations and causes them to exhibit similarities in ecology'. It is recognized that strategies exhibited in the established (adult) phase of a plant's development may differ from those apparent in the regenerative (juvenile) stages (Grime *et al*, 2007). Since the present guide seeks to use mature plants, only the established strategy is referred to in the tables.

Grime *et al.* (2007) ascribe to each plant species a position, or range of positions, within a triangular array referred to as the Competitor (C)–Stress tolerator (S)–Ruderal (R) model of primary ecological strategies (the C-S-R Model). Primary ecological strategies are those related to resource capture, growth and reproduction, and the model is an attempt to recognise the main avenues of ecological specialization in the established, adult phase of the plant (Grime *et al*, 1990).

The C-S-R model proposes that the vegetation which develops in a particular place and at a particular time is the result of an equilibrium which is established between the intensities of stress (constraints on production), disturbance (physical damage to the vegetation), and competition (the attempt by neighbours to capture the same unit of resource). The C-S-R theory hypothesizes that there are three primary strategies in plants, because of these three distinct threats to their existence. According to Grime *et al.* (1990), each threat occurs under particular types of environmental conditions and each confers a selective advantage upon a different type of ecological specialization. Competitors (C) prevail under the threat of competitive exclusion (e.g. in a climax plant community such as an ancient semi-natural woodland where all niches are filled), stress tolerators (S) thrive best under conditions of severe stress (e.g. on a rock face where environmental conditions vary sharply) and ruderals (R) thrive under conditions of severe and frequent disturbance such as in an arable field.

The full spectrum of habitat conditions and associated plant strategies can be represented by an equilateral triangle within which variation in the relative importance of competition, stress and disturbance controls not only three types of plant specialization, but also a range of intermediate strategies (C-R, C-S, S-R and C-S-R), each associated with a less extreme equilibrium between competition, stress and disturbance (Grime *et al.*, 1990). It is important to recognise that the C-S-R equilibrium varies from place to place, even within a plant community, and on diurnal, seasonal and successional timescales. For this reason, communities often contain species of widely differing strategy (Grime *et al.*, 2007).

To aid in understanding the value of the C-S-R model, Grime *et al.* (1990) cite various plant attributes associated with the three ecological strategies as follows:

COMPETITORS

C-strategists monopolize resource capture in productive, relatively undisturbed environments. Attributes associated with the C-strategy include a high potential vegetative growth rate, tall stature and a tendency to form a consolidated growth form by vigorous lateral spread above and below ground, e.g. common nettle (*Urtica dioica)*.

STRESS TOLERATORS

S-tolerators exploit habitats where productivity is low and resource availability brief and unpredictable (e.g. where light occurs as sun flecks under a dense woodland canopy and/or where mineral nutrients become available as short, rich pulses, as from intermittent decomposition events). In such circumstances, conservation of captured resources is of primary significance and success relies upon the capture and retention of scarce resources in a continuously hostile physical environment. S-tolerators therefore demonstrate low potential relative growth rates and delayed onset of reproduction. The leaves of S-tolerators tend to be comparatively long-lived, morphologically conservative structures which are strongly defended against herbivory. In contrast, C-strategists may have more ephemeral leaves allowing them to adjust their canopy shape in accordance with shifting changes in the foliage of their surrounding competitors. Examples of stress tolerators are biting stonecrop (*Sedum acre,* a species of droughted habitats) and sanicle (*Sanicula europaea*, a species of shaded woodland floors where light is limiting).

RUDERALS

R-strategists show a high relative growth rate during the seedling phase and early onset of reproduction (often involving self-pollination and rapid seed production). Arable weeds and ephemeral plants of frequently disturbed habitats such as paths rely on rapid reproduction and dispersal to exploit early successional environments before the vegetation closes (e.g. pineappleweed, *Matricaria discoidea*).

Strategies quoted within Table 1, are taken from Grime *et al.* (2007).

HOW TO USE THE TABLES

The first table (Table of Indicators 1) includes synopses and images for a range of indicator species in the following group order: flowering plants, pteridophytes (ferns and their allies), bryophytes (mosses), algae, fungi (and fungus-like organisms), lichenised fungi (or lichens) and bacteria. In each group, species are listed in alphabetical order by their scientific name. As some species do not have common names, the scientific name has been presented first with the common name (where one exists) in brackets to follow. The images are provided merely to confirm species identity as the present work assumes that the reader will use the tables to seek information for species identified from other sources. In this way, the present work is complementary to a traditional field identification guide and takes the reader the extra step from having identified a species to determining what the presence of that species may indicate about the environment where it was found.
The following flow chart is intended to aid in the use of Table of Indicators 1: Species.

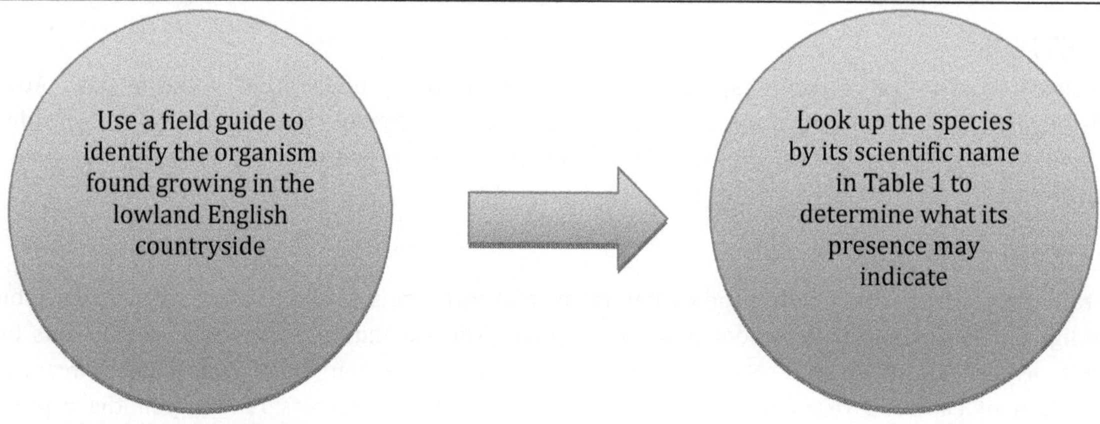

The second table (Table of Indicators 2) is more generic in coverage and introduces the reader to inferences that may be made from communities of plants, or particular habitats and features, which may be found in the countryside. In this regard Table of Indicators 2 provides clues to understanding a range of environmental conditions that may be deduced from the plants present in the vicinity. The conditions are arranged in the following order: soil types, the condition of wetlands, hedgerows and woodlands, man-made disturbance, metal-bearing sites and the age of pastures.

ABBREVIATIONS USED IN THE TABLES

- Ph. 1: Phase 1 habitat code and category, *sensu* JNCC, 1993, 2003
- NVC: National Vegetation Classification community *sensu* Rodwell, 1991 *et seq. cf.* Glossary
- Ell.: Ellenberg Indicator Values *sensu* Hill *et al.*, 1999, 2004, 2007 *cf.* Appendix 1; values shown in bold are considered particularly relevant for the species concerned in the context of the present guide
- Gri.: Ecological strategy *sensu* Grime *et al.*, 2007
- N/L: Species not listed or assigned in the category indicated
- Sdl.: Seedling

TABLE OF INDICATORS 1: SPECIES

Flowering plants

Adoxa moschatellina (moschatel)

Ancient woodland species. Also found in hedgerows and sunken lanes.

Indicator: Ancient woodland

References: Rose (1999)

Ph. 1: A Woodland and scrub, J2 Boundaries;
NVC: W8, 9, 19;
Ell.:

L	F	R	N	S
4	5	6	5	0

Gri.: SR

Ailanthus altissima (tree of Heaven)

A non-native heat-loving street tree from Asia. Invasive in Europe and elsewhere, colonising transport corridors. Suckering and presence of saplings of this species in a town or city may help delineate the extent of the 'urban heat island'.

Suckering indicates: Urban heat island (the artificially warm climate in industrial towns and cities which results from power use)

References: Ellenberg *et al.* (1991),Gilbert (1991), Kowarik & Säumel (2007), Sukopp & Weiler (1986)

Ph. 1: A3 Parkland and scattered trees (or target note e.g. urban boundaries); NVC: N/L;
Ell.:

L	F	R	N	S	T*
8 sdl.	4	x	x	1	**8**

* These are continental values from Kowarik & Säumel (2007), where x indicates indifference. NB Ellenberg T (temperature) values are not considered satisfactory for Britain's climate and ecological requirements of a species in Britain may vary from those of the same species on the continent (Hill *et al.*, 1999).
Gri.: N/L

Ajuga reptans (bugle)

Species of woodlands and open sites but always damp.

Indicator: Dampness, also found in neutral grassland

References: Crawley (2005), Riddelsdell *et al.* (1948)

Ph. 1: A Woodland and scrub, B2.1 Neutral grassland – unimproved, B5 Marsh/marshy grassland;C3.1 Tall ruderal;
NVC: W2, 3, 5, 7-12; M22, 27;MG3; OV26;
Ell.:

L	**F**	R	N	S
5	7	5	5	0

Gri.: R/CSR

Allium ursinum (ramsons, wild garlic)

A species of damp but well-drained deciduous woodland. It is intolerant of waterlogging.

Indicates: Absence of waterlogging

References: Tutin (1957)

Ph. 1: A Woodland and scrub, C3.1 Tall ruderal;
NVC: W8, 9, 12, 21; OV 27;
Ell.:

L	**F**	R	N	S
4	**6**	7	7	0

Gri.: SR

Anaphalis margaritacea (pearly everlasting)

North American introduction, naturalized in UK. Thrives on coal spoil but also other open habitats e.g. by rivers, in grassland. A characteristic species of coal spoil especially in South Wales where it can become sufficiently abundant to colour hillsides

Indicates: Open, disturbed, infertile habitats, especially coal spoil

References: Miller *et al.* (2007), Stace (2010)

Ph. 1: C3.1 Tall ruderal, I2.2 Mine spoil;
NVC: N/L;
Ell.:

L	F	R	N	S
8	5	6	**3**	0

Gri.: N/L

Anemone nemorosa (wood anemone)

Ancient woodland species also found in grassland and dry heath.

Indicator: Ancient semi-natural woodland

Ph. 1: A Woodland and scrub, B1.1 Acid grassland – unimproved, B2.1 Neutral grassland – unimproved, B 3.1 Calcareous grassland – unimproved, B5 Marsh/marshy grassland, C2 Upland species-rich ledges, C3.2 Tall herb and fern – non-ruderal, D1.1 Dry dwarf shrub heath – acid;
NVC: W7-12, 14, 17, 19-21, 25; M26; H12, 16, 18; MG3; CG11, 12, 14; U4, 5, 16, 17;
Ell.:

L	F	R	N	S
5	6	5	4	0

Gri.: SR

Anthriscus sylvestris (cow parsley)

A nitrophile and moist site indicator.

Indicates: Nitrogen rich, moist site

References: Hill *et al.* (1999), Pitcairn *et al.* (2006)

Ph. 1: A Woodland and scrub, B2 Neutral grassland, C3.1 Tall ruderal, F Swamp, marginal/inundation, H2 Saltmarsh;
NVC: W8-10, W12, W24; MG1-MG3; S17; SM28; OV24-27;
Ell.:

L	F	R	N	S
6	**5**	7	7	0

Gri.: C/CR

Apium nodiflorum (fool's watercress)

A species of fertile wet sites where the growth of taller species is restricted by disturbance such as erosion or ditch clearance. It is a palatable species which can be eliminated by grazing.

Indicates: Fertile, wet, disturbed sites with little grazing pressure

References: Grime *et al.* (2007), Preston & Croft (1997)

Ph. 1: A Woodland and scrub, F2 Marginal and inundation, G1 Standing water;
NVC: W1, 6, A1, 2, 8, 19; S12, 14, 15, 22, 23, 26, 27;
Ell.:

L	F	R	N	S
7	**10**	7	7	0

Gri.: C-R

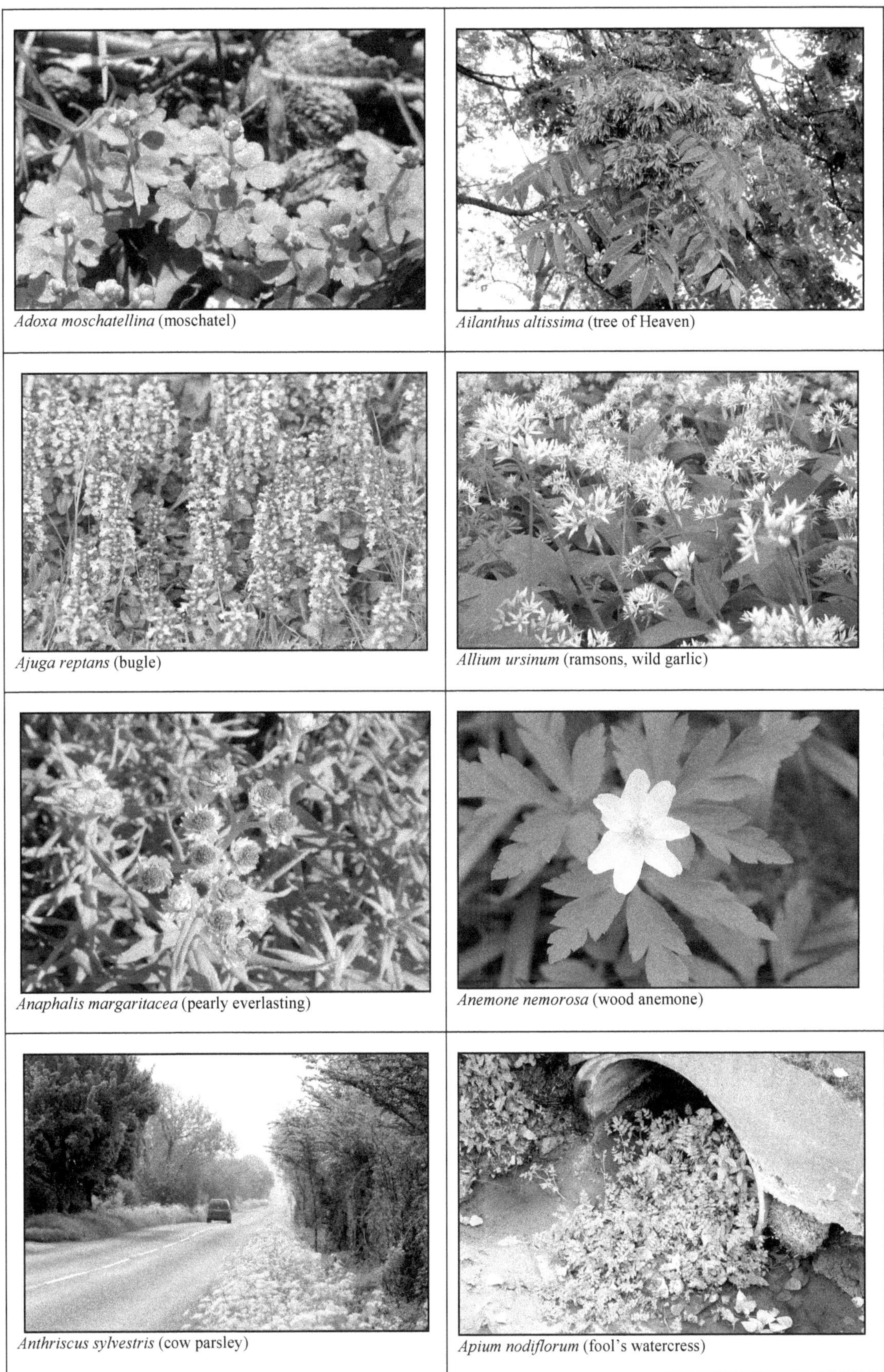

Adoxa moschatellina (moschatel)

Ailanthus altissima (tree of Heaven)

Ajuga reptans (bugle)

Allium ursinum (ramsons, wild garlic)

Anaphalis margaritacea (pearly everlasting)

Anemone nemorosa (wood anemone)

Anthriscus sylvestris (cow parsley)

Apium nodiflorum (fool's watercress)

Armeria maritima (sea thrift)

Usually coastal but inland, it is largely confined to metal-rich soils or salted road verges. Look for other halophytes (maritime plants able to exploit the stressed, road-edge environment subject to winter salting, exhaust and road run-off). Species include: *Bassia scoparia, Cochlearia danica,Dezmazeria marina, Plantago coronopus, Plantago maritima, Puccinellia distans.*

Indicator: Metallophyte/halophyte zone/winter salting (if inland)

References: DoE (1994), Dowdeswell (1987), Rich (2001), Richards *et al.* (1989), Scott (1985), Scott & Davison (1982)

Ph. 1: A2 Scrub, B3.1 Calcareous grassland (unimproved), C3 Upland species-rich ledges, D1.1 Dry dwarf shrub heath – acid, D3 Lichen/bryophyte heath, D4 Montane heath/dwarf herb, E2.2 Basic flush, F2.2 Marginal, H2 Saltmarsh, H4 Rocks/boulders above high-tide mark, H5 Strandline vegetation, H8.3 Crevice/ledge vegetation, H8.4 Coastal grassland, H8.5 Coastal heathland, J2.5 Wall;
NVC: W22; M12; H7, 19, 20; CG1, 11, 12, 14; U9, 10, 13, 14, 17; S4; SM10, 13-22, 24, 26; SD3; MC1-6, 8-12; OV41;
Ell.:

L	F	R	N	**S**
8	7	5	5	**3**

Gri.: N/L

Arrhenatherum elatius (false oat-grass)

Grassland with conspicuous false oat-grass and tall umbellifers (e.g. cow parsley) is an ungrazed / irregularly grazed plant community exposed to some nutrients e.g. run-off. If totally ungrazed, there is an annual cut e.g. as on road verges.

Indicates: Lack of grazing

References: Rodwell (1992)

Ph. 1: A Woodland and scrub, B2 Neutral grassland, B3 Calcareous grassland, B4 Improved grassland, B5 Marsh/marshy grassland, C3 Tall herb and fern, E3.2 Fen-basin mire, F Swamp, marginal and inundation, H3 Shingle/gravel above high-tide mark, H2 Saltmarsh, H6.4 Dune slack, H6.5 Dune grassland, H6.7 Dune scrub, H8.4 Coastal grassland, H8.4 Coastal grassland, J1 Cultivated/disturbed land, J2.5 Wall;
NVC: W6, 8-10, 12, 21, 23-25; M22, 27, 28; MG1-7, 9, 11-12; CG2-4, 6-8; U16; S7, 25, 26; SM28; SD1, 2, 7-9, 15, 18; MC12; OV 9, 18, 22-27, 30, 37-42;
Ell.:

L	F	R	**N**	S
7	5	7	7	0

Gri.: C (*bulbosum*)/CR

Arum maculatum (lords-and-ladies, cuckoo-pint)

Grows on many kinds of soil, provided that they are rich in bases (somewhat alkaline with a good quantity of elements like calcium). Most abundant on calcareous soils and absent from acid, peaty, base-deficient soils.

Indicates: Base rich soil

References: Prime (1981)

Ph. 1: A Woodland and scrub;
NVC: W8, 9, 12-14, 21, 24, 25;
Ell.:

L	F	**R**	N	S
4	5	7	7	0

Gri.: SR

Asperula cynanchica (squinancywort)

A species of calcareous grassland requiring a high soil pH and calcium level.

Indicator: Calcicole and a dry site indicator

References: Hill *et al.* (2004), Steele (1955)

Ph. 1: B3 Calcareous grassland;
NVC: CG1-7, 9;
Ell.:

L	**F**	**R**	N	S
7	**3**	**8**	2	0

Gri.: N/L

Aster novae-belgii group (Michaelmas daisies)

Garden escapes found in disturbed sites and waste ground. *A. novae-belgii* now known to be a group of several similar species. Can be useful in identifying former railway halts (defunct flower beds) on abandoned railway lines.

Indicates: Disturbance

References: Crawley (2005)

Ph. 1: C3.1 Tall ruderal;
NVC: N/L;
Ell.:

L	F	R	N	S
7	6	7	6	1

Gri.: C/CSR

Bellis perennis (daisy)

A species of mown, heavily grazed or trampled, neutral or calcareous grassland. Does best in grasslands that are relatively wet for at least part of the year.

Indicates: Sward height is controlled

References: Preston *et al.* (2002)

Ph. 1: B1 Acid grassland, B2.1 Neutral grassland – unimproved, B2.2 Neutral grassland – semi-improved, B3 Calcareous grassland, B4 Improved grassland, B5 Marsh/marshy grassland, E2.3 Bryophyte dominated spring, F2.2 Inundation, H2 Saltmarsh, H6.4 Dune slack, H6.5 Dune grassland, H8.3 Crevice/ledge vegetation, H8.4 Coastal grassland, J1 Cultivated disturbed land;
NVC: M32, 38; MG3-8, 11; CG1-10, 13, 14; U4; SD8, 9, 14, 16, 17; MC5, 9-11; OV4, 6, 21-23;
Ell.:

L	F	R	N	S
8	5	6	4	0

Gri.: R/CSR

Berberis darwinii (Darwin's berberry)

Spiny and often planted to discourage access to or across flower beds in urban sites around offices, car parks or schools. May spread by seed to other sites.

Indicates: Former urban planting (see also *Pyracantha*)

References: Crawley (2005)

Ph. 1: J1.4 Introduced shrub, J2 Boundary;
NVC: N/L;
Ell.: N/L;
Gri.: N/L

Berberis vulgaris (barberry)

The alternative host to wheat in the life cycle of the crop disease wheat rust (*Puccinia graminis*), therefore barberry was eradicated from many hedgerows in the past.

Indicates: Persistence in a rural hedge might indicate past agricultural practices (i.e. **grazing rather than cereal growing)**

References: Preston *et al.* (2002), Webster & Weber (2007)

Ph. 1: J2 Boundary;
NVC: N/L;
Ell.:

L	F	R	N	S
7	4	8	3	0

Gri.: N/L

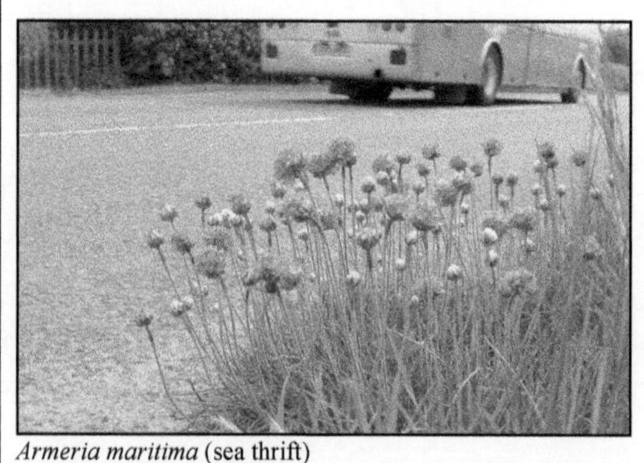 *Armeria maritima* (sea thrift)	 *Arrhenatherum elatius* (false oat-grass)
 *Arum maculatum* (lords-and-ladies, cuckoo-pint)	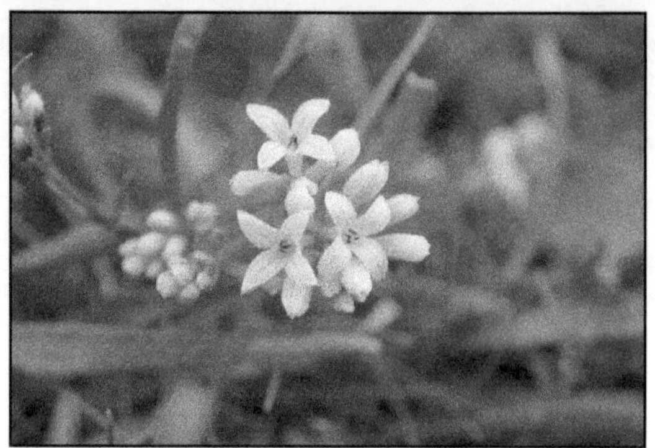 *Asperula cynanchica* (squinancywort)
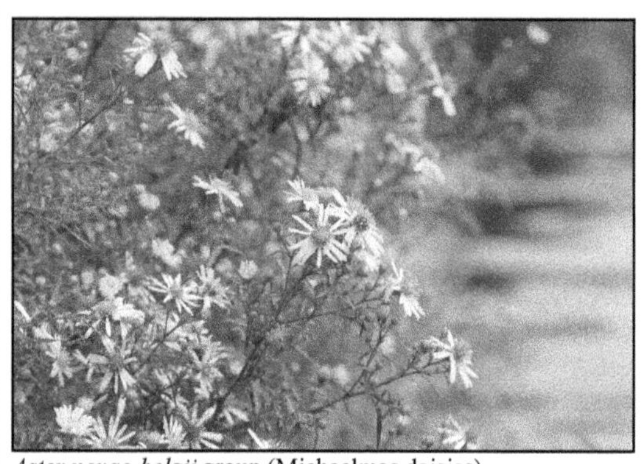 *Aster novae-belgii* group (Michaelmas daisies)	 *Bellis perennis* (daisy)
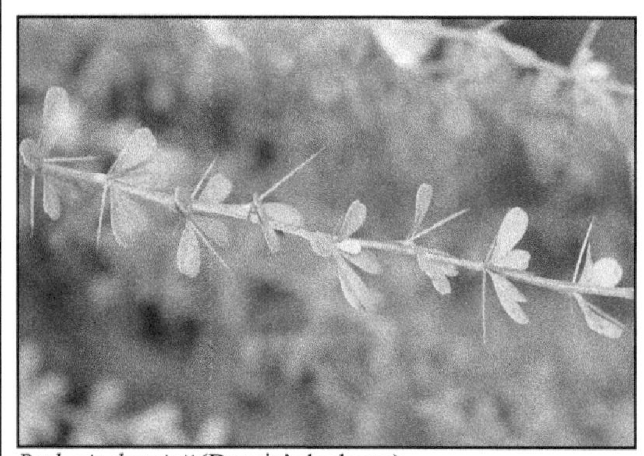 *Berberis darwinii* (Darwin's berberry)	 *Berberis vulgaris* (barberry)

Blackstonia perfoliata (yellow-wort)

A species of calcareous grasslands and dunes.

Indicates: Calcareous soils

Ph. 1: B3 Calcareous grassland, H6.4 Dune slack, H8.4 Coastal grassland;
NVC: CG1-5, 7, SD13, 14, MC11;
Ell.:

L	F	**R**	**N**	S
8	5	**8**	**2**	0

Gri.: SR

Brachypodium pinnatum (tor grass) (includes *B. rupestre*)

Its spread has been linked with nitrogen enrichment in calcareous grassland, this associated with a reduction in botanical diversity, but results are conflicting.

Indicates: Spread may indicate nutrient enrichment of grassland

References: Bobbink & Lamers (2002) for a discussion

Ph. 1: A2 Scrub, B3 Calcareous grassland, H8.3 Crevice/ledge vegetation, H8.4 Coastal grassland;
NVC: W21; CG2, 3, 4, 5; MC4, 5, 8, 11;
Ell.:

L	F	**R**	N	S
7	3	**8**	**3**	0

Gri.: SC

Buddleja davidii (butterfly bush)

A horticultural escape. Found in dry, disturbed areas. Prolific seed production and wide environmental tolerance enable it to colonize brownfield sites. Large populations can develop from the wind-dispersed seed.

Indicates: Disturbance

References:Preston *et al.* (2002), Tallent-Halsell & Watt (2009)

Ph. 1: C3.1 Tall ruderal, J1.2 Cultivated/disturbed land – amenity grassland (as seedlings), J1.4 Introduced shrub, J2.5 Wall;
NVC: OV23, 24;
Ell.:

L	F	R	N	S
7	5	7	5	0

Gri.: CL

Callitriche spp. (star-worts)

Most star-worts are pioneers, which colonize disturbed wetland substrates, later disappearing as competition intensifies.

Indicates: Disturbance

References: Schotsman (1972)

Ph. 1: A Woodland and scrub, E2.2 Basic flush, E3.2 Basin mire, F Swamp, marginal and inundation, G Open water;
NVC: W1, 3, 5, 8; M35; A1-6, 8-17, 19-21, 23, 24; S4-6, 10, 12-14, 16, 23, 27, 28; OV30-32, 35;
Ell.: For *C. stagnalis* (values vary for other species):

L	**F**	R	N	S
7	**10**	6	6	1

Gri. (*C. stagnalis*): R/CR

Calluna vulgaris (heather)

Heather will not grow on calcareous soils. It is a characteristic species of heathland. If found in a chalk or limestone area, it indicates the presence of a pocket of non-calcareous soil. This may be due to leaching of bases from the profile or superficial deposits of non-calcareous material.

Indicates: Acid soil (a calcifuge)

References: Rayner (1921)

Ph. 1: A Woodland and scrub, B1 Acid grassland, B2.1 Neutral grassland – unimproved, B3.1 Calcareous grassland – unimproved, B5 Marsh/marshy grassland, C1 Bracken, C2 Upland species-rich ledges, C3 Tall herb and fern, D Heathland (various), E Mire (various), F2.2 Inundation, H6.5 Dune grassland, H6.6 Dune heath, H8.4 Coastal grassland;
NVC: W4, 11, 15-20, 23; M2, 8, 11, 13-21, 24, 25; H1-22; MG5; CG9-14; U1-7, 15-17, 19-21; SD11, 12; MC10; OV27, 34, 37;
Ell.:

L	F	**R**	**N**	S
7	6	**2**	**2**	0

Gri.: S/SC

Campanula rotundifolia (harebell)

A species of infertile habitats but tolerant of a wide range of soil pH. A nitrophobe (indicator of low nitrogen).

Indicates: Nutrient-poor soil

References: Pitcairn *et al.* (2006), Preston *et al.* (2002)

Ph. 1: A Woodland and scrub, B Grassland (unimproved), C1 Bracken (understory), C2 Upland species rich ledges, C3.2 Tall herb and fern – non-ruderal, D Heathland, E2.2 Basic flush, H6.5 Dune grassland, H6.6 Dune heath, H8.4 Coastal grassland, I Rock exposure and waste;
NVC: W11, 19, 20, 23; M11; H1, 10-12, 16, 18; MG1-3; CG1-14; U1, 4, 5, 10, 13-17, 19, 20; SD 7-9, 11, 12; MC9; OV37-40;
Ell.:

L	F	R	**N**	S
7	4	5	**2**	0

Gri.: S/CSR

Cardamine pratensis (lady's smock, cuckoo flower)

An Ellenberg moisture value of 8 indicates that *Cardamine pratensis* is a conspicuous indicator of very moist sites.

Indicates: Wet grassland

References: Hill *et al.* (2004)

Ph. 1: A Woodland and scrub, B1.1 Acid grassland – unimproved, B2 Neutral grassland, B3.1 Calcareous grassland – unimproved, B4 Improved grassland, B5 Marsh/marshy grassland, C3.1 Tall ruderal, E2.1 Acid/neutral flush, E2.2 Basic flush, E2.3 Bryophyte dominated spring, E3.2 Basin mire, F Swamp, Marginal and inundation; H6.4 Dune slack;
NVC: W3, 5, 7; M4-6, 8-11, 13, 22-25, 27, 28, 32, 35, 37, 38; MG3-10; CG10, 12; U6; S1, 3, 4, 6, 9, 18, 19, 24, 27; SD15, 17; OV26, 35;
Ell.:

L	**F**	R	N	S
7	**8**	5	4	0

Gri.: R/CSR

Carex flacca (glaucous sedge)

A neutral-calcareous grassland nitrophobe, the persistence of which is reduced if nitrogen is added

Indicates: Low soil nitrogen(an unimproved site)

References: Pitcairn *et al.* (2006)

Ph. 1: A Woodland and scrub, B2.1 Neutral grassland – unimproved, B3.1 Calcareous grassland – unimproved, B5 Marsh/marshy grassland, C2 Upland species rich ledges, D1 Dry dwarf shrub heath – acid/basic, D2 Wet dwarf shrub heath, D4 Montane heath/dwarf herb, D5 Dry heath/acid grassland, E2.2. Basic flush, E2.3 Bryophyte dominated spring, E3.2 Basin mire, F2.2 Marginal/inundation – inundation, H6.4 Dune slack, H6.5 Dune grassland, H8.3 Crevice/ledge vegetation, H8.4 Coastal grassland, I1.2.2 Scree-basic, I1.3 Limestone pavement;
NVC: W12, 19, 20; M9-11, 13, 22, 25, 26, 37, 38; H4-8; MG5, 8; CG1-11, 13, 14; U15, 17; SM16; SD8, 9, 13-17; MC5, 9, 11; OV34, 37, 38;
Ell.:

L	F	R	**N**	S
7	5	6	**2**	0

Gri.: S

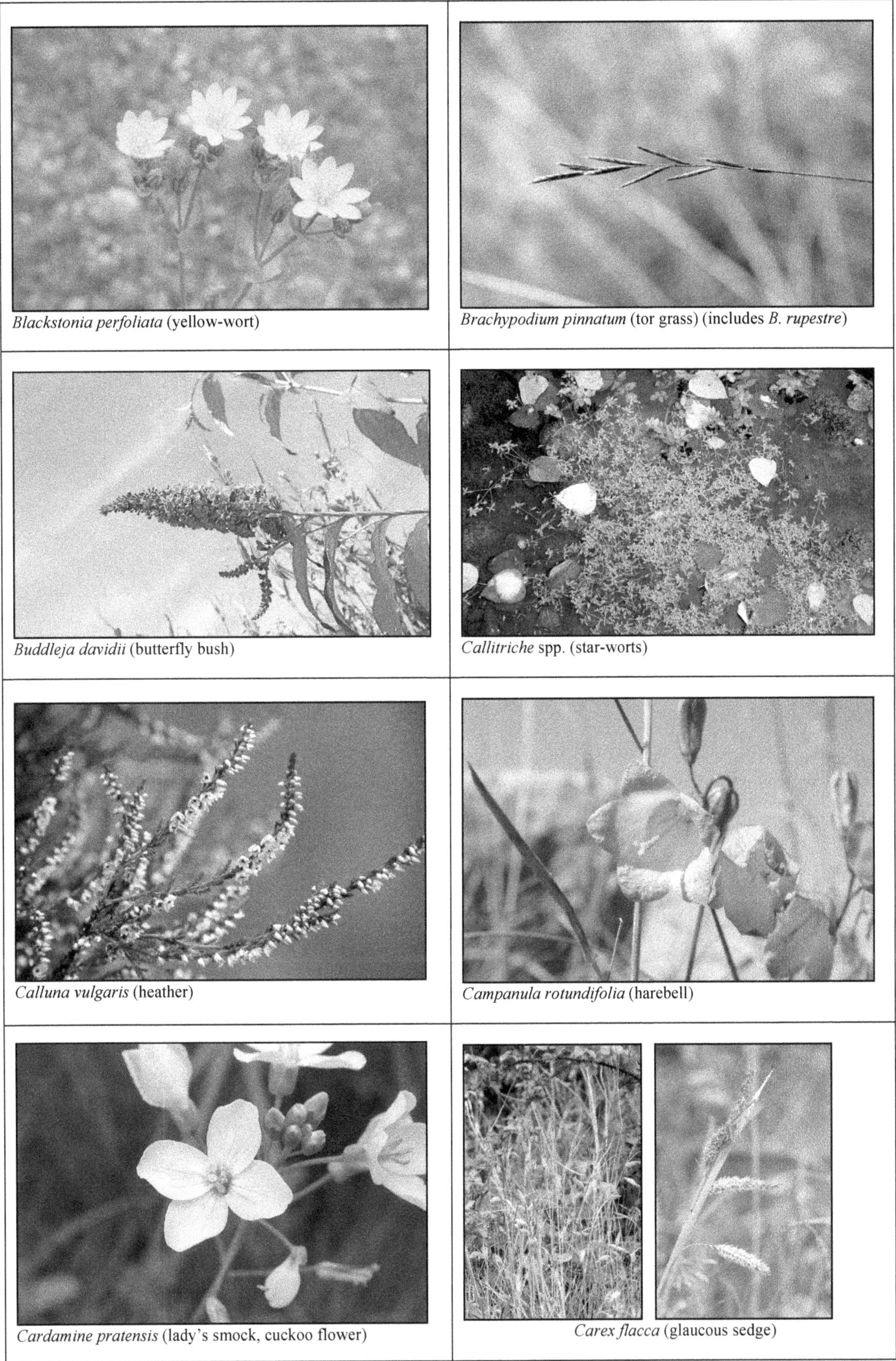

Blackstonia perfoliata (yellow-wort)

Brachypodium pinnatum (tor grass) (includes *B. rupestre*)

Buddleja davidii (butterfly bush)

Callitriche spp. (star-worts)

Calluna vulgaris (heather)

Campanula rotundifolia (harebell)

Cardamine pratensis (lady's smock, cuckoo flower)

Carex flacca (glaucous sedge)

Carex pendula (pendulous sedge)

A species of ancient semi-natural woodland (but sometimes planted).

Indicator: Ancient semi-natural woodland

References: Rose (1999)

Ph. 1: A Woodland and scrub; NVC: W7, 8;
Ell.:

L	F	R	N	S
5	8	7	6	0

Gri.: C/SC

Carlina vulgaris (carline thistle) (shown with *Calluna vulgaris)*

Shows a preference for calcareous soils but sometimes found on others.Behaves as a calcicole in southern England but almost indifferent to soil composition on the continent. Also associated with coal waste when it may grow with calcifuge species such as *Calluna vulgaris.*

Indicates: Often on calcareous sites, also associated with coal waste

References: Kay (1994), Lousley (1969), Miller *et al.* (2007), Steele (1955)

Ph. 1: B3.1 Calcareous grassland, C3.1 Tall ruderal, H6.4 Dune slack, H8.4 Coastal grassland, I1.1.2 Inland cliff – basic, I2 Artificial exposures (and waste tips), J2.5 Wall;
NVC: CG1-5, 7-9; SD16; MC11; OV27, 39, 41;
Ell.:

L	F	**R**	**N**	S
8	4	**7**	**2**	0

Gri.: SR

Chamerion angustifolium (Rose-bay willowherb, 'fireweed')

Proliferated on bomb sites in World War II. Tolerates a wide range of soils, but common on disturbed ground. Colonizes rapidly from airborne seeds, sprouts from rhizomes following disturbance and is an important colonizer after fire. It has some potential to establish on areas contaminated with crude oil.

Indicates: Disturbance and/or burning

References: Grigson (1958), Hendrickson (1972), Kershaw & Kershaw (1986), Pavek (1992), Stickney (1986, 1990)

Ph. 1: C3.1 Tall ruderal, F2.2 Inundation, H6.5 Dune grassland, H6.7 Dune scrub, H6.8 Open dune, J1.1 Arable, J1.2 Amenity grassland, J1.3 Ephemeral/short perennial;
NVC: SD 5-7, 10, 18; OV9, 10, 19, 21-23, 27, 32;
Ell.:

L	F	R	N	S
6	5	6	5	0

Gri.: C

Cirsium arvense (creeping thistle)

An indicator of good agricultural land and fertile soil including gardens. Its growth is stimulated by fertilizers and it does not occur on very acid soils. A plant of open sites.

Indicates: Fertile soil

References: Fairbairn & Thomas (1959), Tiley (2010)

Ph. 1: A2 Scrub, B1.2 Acid grassland – semi-improved, B2 Neutral grassland, B3 Calcareous grassland, B4 Improved grassland, B5 Marsh/marshy grassland, C1 Bracken, C3.1 Tall ruderal, E3.2 Basin mire, F Swamp, marginal and inundation, H Saltmarsh, H3 Shingle/gravel above high tide mark, H4 Rocks/boulders above high-tide mark, H5 Strandline vegetation, H6.4 Dune slack, H6.5 Dune grassland, H6.7 Dune scrub, H6.8 Open dune, H8.3 Crevice/ledge vegetation, H8.4 Coastal grassland, J1 Cultivated/disturbed land, J2.5 Wall;
NVC: W21, 24, 25; M22, 24, 27, 28; MG1, 4 -7, 9-13; CG 2-4, 6; U4, 20; S5, 6, 17, 18, 21, 23, 24, 26, 28; SM 18, 28; SD1, 2, 4-9, 12, 15, 16, 18, 19; MC4, 8; OV3-5, 7-10, 13-16, 19-29, 31-33, 41;
Ell.:

L	F	R	N	S
8	6	7	6	0

Gri.: C

Cirsium eriophorum (woolly thistle)

A species of dry, calcareous grassland. In Britain, the species is restricted to freely draining base-rich soils over chalk and limestone.

Indicator: Calcareous substrate

References: Tofts (1999)

Ph. 1: B3 Calcareous grassland;
NVC: CG2, 3, 5;
Ell.:

L	**F**	**R**	N	S
8	**4**	**8**	5	0

Gri.: N/L

Cirsium vulgare (spear thistle)

A species of nitrogen-rich soil.

Indicates: Nitrogen-rich soil

References: Ellenberg (1979), Klinkhamer & de Jong (1993)

Ph. 1: A2 Scrub, B1.2 Acid grassland – semi-improved, B2.2 Neutral grassland – semi-improved, B3 Calcareous grassland, B4 Improved grassland, C1 Bracken, C3.1 Tall ruderal, F2.2. Inundation, H2 Saltmarsh, H3 Shingle/gravel above high-tide mark, H6.5 Dune grassland, H6.7 Dune scrub, H6.8 Open dune, H8.4 Coastal grassland, J1 Cultivated/disturbed land;
NVC: W24; MG1, 6, 7, 9, 11; CG2, 4, 6-9; U4, 20; SM28; SD1, 4, 6-8, 18; MC11, 12; OV4, 10, 13, 15, 19, 22, 23, 25, 27, 33, 37;
Ell.:

L	F	R	N	S
7	5	6	6	0

Gri.: CR

Clematis vitalba (old man's beard, traveller's joy)

An indicator of calcareous soil.

Indicates: Calcareous soils

References: Lousley (1969)

Ph. 1: A Woodland and scrub, B3 Calcareous grassland;
NVC: W8, 12, 13, 21; CG3, 5;
Ell.:

L	F	**R**	N	S
6	4	**8**	5	0

Gri.: SC

Cochlearia danica (Danish scurvy-grass)

A coastal species, found inland on roadsides and central reservation as a member of the roadside halophyte zone, possibly as a result of winter salting activity.

Indicates: Halophyte zone

References: Scott & Davison (1982), Smith & Sangwine (2002)

Ph. 1: H2 Saltmarsh, H4 Rocks/boulders above high-tide mark, H8.3 Crevice/ledge vegetation, H8.4 Coastal grassland;
NVC: SM21, 25; MC1, 5, 8, 12;
Ell.:

L	F	R	N	**S**
9	6	7	5	**4**

Gri.: SR

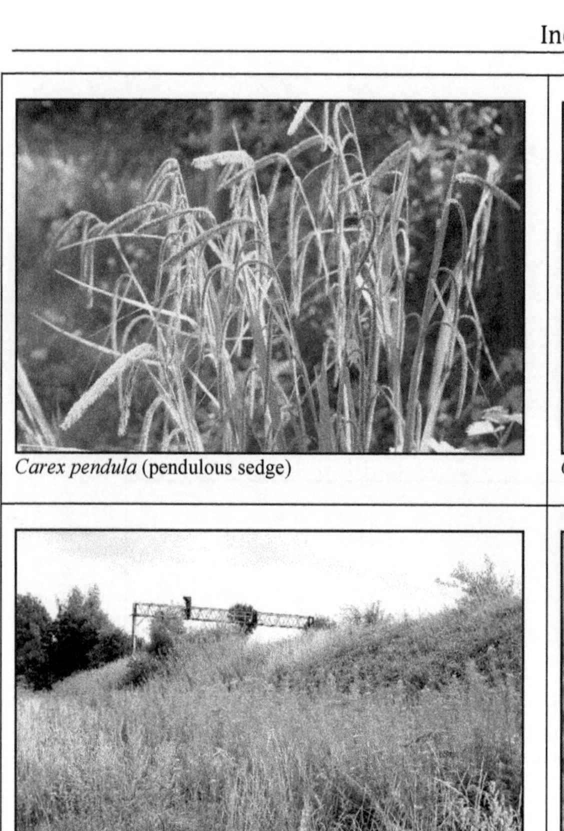 *Carex pendula* (pendulous sedge)	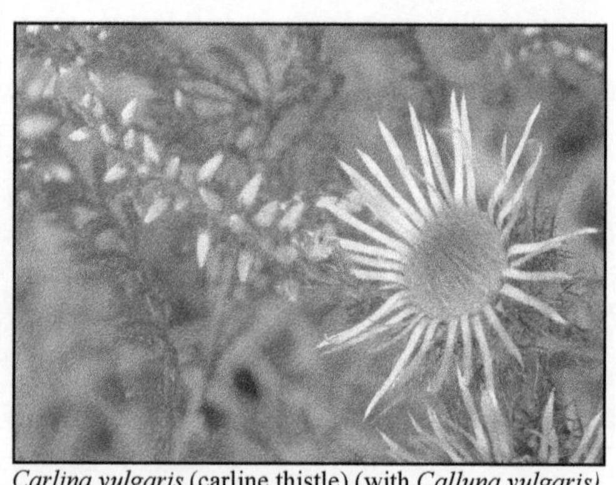 *Carlina vulgaris* (carline thistle) (with *Calluna vulgaris)*
 *Chamerion angustifolium* (Rose-bay willowherb, 'fireweed')	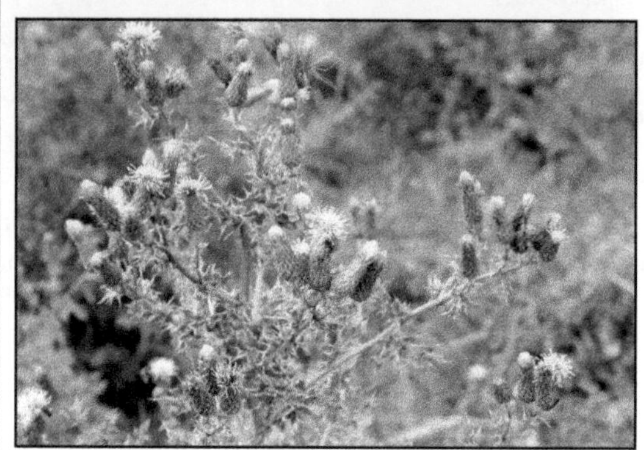 *Cirsium arvense* (creeping thistle)
 *Cirsium eriophorum* (woolly thistle)	 *Cirsium vulgare* (spear thistle)
 *Clematis vitalba* (old man's beard, traveller's joy)	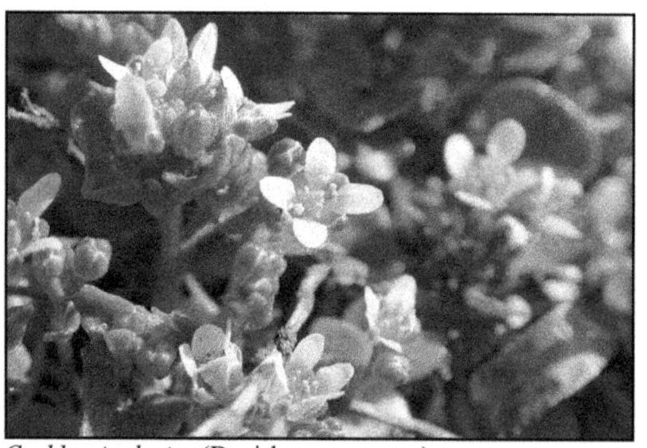 *Cochlearia danica* (Danish scurvy-grass)

Colchicum autumnale (meadow saffron)

Not usually tolerated in grazing meadows because the plant is poisonous to cattle. Associated with ancient semi-natural woodland.

Indicates: Damp, grassy places where cattle are not usually grazed or ancient semi-natural woodland

References:Crawley (2005), Mabey (1996), Rose (1999)

Ph. 1: A1.1.1 Broadleaved woodland – semi-natural; B2 Neutral grassland – unimproved;
NVC: MG5;
Ell.:

L	F	R	N	S
6	6	6	4	0

Gri.: N/L

Conium maculatum (hemlock)

Common in disturbed habitats in the British Isles. Also near rabbit warrens as it is apparently not grazed by rabbits.

Indicates: Disturbed sites

References: Salisbury (1961), Tremlett *et al.* (1984)

Ph. 1: C3.1 Tall ruderal, F2.1 Marginal, J1.3 Ephemeral/short perennial; NVC: S23; OV22, 24, 25;
Ell.:

L	F	R	**N**	S
8	5	7	**8**	0

Gri.: C/CR

Conopodium majus (pignut)

A species of woodland and/or grassland which has existed for some significant time.

Indicates: Continuity of habitat

References: Rackham (2006)

Ph. 1: A Woodland and scrub, B1 Acid grassland, B2.1 Neutral grassland – unimproved, C1 Bracken, C3.2 Other tall herb and fern – non-ruderal, H8.4 Coastal grassland;
NVC: W7-11, 25; MG1, 3, 5, 9; U4, 16, 20; MC9;
Ell.:

L	F	R	N	S
6	5	5	5	0

Gri.: SR

Cornus sanguinea (dogwood)

Presence in quantity with privet (*Ligustrum vulgare*) and wayfaring tree (*Viburnum lantana*) indicates a calcareous soil. Often included in landscape planting.

Indicates: Calcareous soil

References: Lousley (1969)

Ph. 1: A Woodland and scrub, J2 Boundaries;
NVC: W8, W12, W21;
Ell.:

L	F	**R**	N	S
7	5	**7**	6	0

Gri.: SC

Crataegus laevigata (Midland hawthorn) – two fruits shown dissected

Associated with ancient hedges and ancient woodland. It tends to flower earlier than *C. monogyna* and therefore early flowering hawthorn shrubs in a hedgemay indicate a boundary of some antiquity or ancient woodland remnant (check for species).

Indicates: Ancient hedgerow or woodland

References: Rackham (1986), Rich & Jermy (1998)

Ph. 1: A1 Woodland and scrub, J2 Boundaries;
NVC: W5, 8, 10, 21;
Ell.:

L	F	R	N	S
5	5	7	5	0

Gri.: N/L

Cuscuta epithymum (dodder)

A holoparasitic indicator species of dry European heathland. Parasitic on a range of hosts in acidic heathland habitats including *Ulex* and *Calluna*. It appears in young, recently managed heath vegetation and gradually disappears once *C. vulgaris* becomes older than c. 7 years. Also occurs in habitats with a strong calcareous substrate. In Europe it is strongly connected to oligotrophic soils, both acid and calcareous.

Indicator: Dry heathland

References: Hill *et al.* (2004), Meulebrouck (2009), Schaminée *et al.* (1996), Stace (2010)

Ph. 1: D1.1 Dry dwarf shrub heath (acid), D5 Dry heath/acid grassland mosaic;
NVC: H2-4;
Ell.:

L	F	R	N	S
7	6	2	2	0

Gri.: N/L

Cynosurus cristatus (crested dog's-tail)

Found on a range of soil types but MG5 *Cynosurus cristatus–Centaurea nigra* grassland is typical of traditionally managed, grazed hay meadows on neutral lowland soils ('old meadows') and, where moist and freely draining, MG6 *Lolium perenne–Cynosurus cristatus* grassland is the major permanent pasture type on such soils.

Indicates: A notable component of neutral grassland

References: Rodwell (1992)

Ph. 1: B1 Acid grassland, B2 Neutral grassland, B3 Calcareous grassland, B5 Marsh/marshy grassland, D4 Montane heath/dwarf herb, H2 Saltmarsh, H6.4 Dune slack, H6.5 Dune grassland, H8.4 Coastal grassland, J1.2 Amenity grassland, J1.3 Ephemeral/short perennial;
NVC: M22, 23; MG1, 3; CG1-4, 6, 8-10, 13; U4-6, 8-10, 12; SM16; SD8, 16, 17; MC11; OV21, 23;
Ell.:

L	F	R	N	S
7	5	6	4	0

Gri.: R/CSR

Dactylorhiza fuchsii (common spotted-orchid) or other *Dactylorhiza* spp.

European orchids have complex, mycorrhizal life cycles and long development times. They may therefore be sensitive to disturbance. The development of most species of *Dactylorhiza* from germination to shoot/tuber formation is c. 4 years.

Indicates: Habitat has been suitable for orchid development for a minimum of 4 years

References: Foley & Clarke (2005), Rasmussen (1995)

D. fuchsii: Ph. 1: A2 Scrub, B2 Neutral grassland, B3 Calcareous grassland, B5 Marsh/marshy grassland, C3.1 Tall ruderal, E2.2 Basic flush, E3.2 Basin mire, H6.4 Dune slack;
NVC: W3; M9, 10, 13, 22, 24; MG3, 9; CG2, 3; SD17; OV26;
Ell.:

L	**F**	R	N	S
7	**8**	7	3	0

Gri.: SR

Colchicum autumnale (meadow saffron)

Conium maculatum (hemlock)

Conopodium majus (pignut)

Cornus sanguinea (dogwood)

Crataegus laevigata (Midland hawthorn) – two fruits dissected

Cuscuta epithymum (dodder)

Cynosurus cristatus (crested dog's-tail)

Dactylorhiza fuchsia (common spotted-orchid)

Dactylorhiza fuchsii (common spotted-orchid)

Daphne laureola (spurge laurel)

Associated with ancient woodland but introduced in some sites, e.g. pheasant rearing estates (cover) and parklands.

Indicates: An ancient hedge line or woodland (where not introduced)

References: Preston *et al.* (2002), Rose (1999)

Ph. 1: A1 Woodland, J2 Boundaries;
NVC: W8, 12;
Ell.:

L	F	R	N	S
4	5	7	5	0

Gri.: SC

Daucus carota (wild carrot)

Presence on a road or motorway verge is a useful 'drive-by' marker for an area of species-rich grassland. Earlier in the year, *Leucanthemum vulgare* (Ox-eye daisy) fulfils the same function.

Indicates: Species-rich road verge

Ph. 1: B2 Neutral grassland, B3 Calcareous grassland, C3.1 Tall ruderal, D1 Dry dwarf shrub heath, H6.5 Dune grassland, H6.7 Dune scrub, H8.3 Crevice/ledge vegetation, H8.4 Coastal grassland, H8.5 Coastal heathland, J1.1 Arable, J1.2 Amenity grassland, J2.5 Wall;
NVC: H6-8; MG1; CG1-4, 7, 8, 13; SD8, 18; MC1, 4, 5, 6, 8, 9, 10, 11, 12; OV2, 4, 11, 23, 25, 41;
Ell.:

L	F	R	N	S
8	4	7	3	2

Gri.: SR/CSR

Deschampsia cespitosa (tufted hair-grass)

Grows in a diverse range of soil types but frequently associated with impeded drainage, seasonal flooding and nutrient poor soils.

Indicates: Impeded drainage

References: Davy (1980)

Ph. 1: A Woodland and scrub, B1 Acid grassland, B2 Neutral grassland, B3 Calcareous grassland, B5 Marsh/marshy grassland, C2 Upland species-rich ledges, C3.1 Tall ruderal, C3.2 Tall herb and fern – non-ruderal, D1.1 Dry dwarf shrub heath – acid, D4 Montane heath/dwarf herb, D5 Dry heath/acid grassland, E2 Flush and spring, E3.2 Basin mire, F Swamp, marginal and Inundation, H2 Saltmarsh;
NVC: W2-12, 14, 16, 17, 19, 20, 24; M7, 8, 10-12, 22-26, 28, 31-34, 37, 38; H18, 20; MG2-4, 6, 9, 13; CG2, 9-12, 14; U4, 5, 7-19; S11, 23, 28; SM28, OV26, 27, 32;
Ell.:

L	F	R	N	S
6	6	5	4	0

Gri.: SC/CSR

Digitalis purpurea (foxglove)

A species of acidic soils. Where present in limestone areas, it indicates locally acidic soil conditions, e.g. outcrop of sand or deeper, rain-leached substrate. Often abundant in disturbed or burnt areas, e.g. recently felled forestry plantation. Germination of the seed may be associated with soil disturbance.

Indicator: Calcifuge ('lime hating' species), acidic soils

References: Grime (1979), Preston *et al.* (2002), Zohlen & Tyler (2004)

Ph. 1: A Woodland and scrub, C1 Bracken, C3.1 Tall herb and fern – non-ruderal, C3.2 Tall ruderal, D1 Dry dwarf shrub heath, D5 Dry heath/acid grassland;
NVC: W6-11, 14, 16, 17, 22-25; H8; U16, 20, 21; OV27;
Ell.:

L	F	**R**	N	S
6	6	**4**	5	0

Gri.: SR/CSR

Drosera rotundifolia (round-leaved sundew) and other *Drosera* species

Nitrophobe species of wet acid mire communities.

Indicator: Low nitrogen, acidic wetland

References: Pitcairn *et al.* (2006)

Ph.1 (UK spp. amalgamated): B5 Marsh/marshy grassland, D2 Wet dwarf shrub heath, D6 Wet heath/acid grassland, E Mire;
NVC: M1-4, 6, 8-10, 13-19, 21, 24, 29; H5;
Ell. (*D. rotundifolia*):

L	F	**R**	**N**	S
8	9	**2**	**1**	0

Gr.: SR

Erica cinerea (bell heather)

Characteristic dry heathland species. Calcifuge.

Indicates: Drier heath, acid soils

Ph. 1: A Woodland, B1.1 Acid grassland – unimproved, C1 Bracken, C3.1 Tall ruderal, C3.2 Tall herb and fern – non-ruderal, D Heathland, E 1.6.1 Blanket bog, E1.6.2 Raised bog, E1.7 Wet modified bog, E1.8 Dry modified bog, F2.2. Inundation, H6.5 Dune grassland, H6.6 Dune heath;
NVC: W11, 16-18; M15, 17; H1-8, 10-17, 20, 21; U3, 5, 16, 19-21; SD12; OV27, 34;
Ell.:

L	**F**	**R**	N	S
7	**5**	**2**	2	0

Gri.: S/SC

Erica tetralix (cross-leaved heath)

Characteristic of wet heath and mire communities and regarded as an indicator of wet heath.

Indicates: wet heath, acid soils

References: Bannister (1966), Box *et al.* (2011)

Ph.1: A Woodland and scrub, B1.1 Acid grassland – unimproved, B3.1 Calcareous grassland – unimproved, D Heathland, E1 Bog, E2.1 Acid/neutral flush, E2.2 Basic flush, B5 Marsh/marshy grassland, E3.1 Valley mire, H8.4 Coastal grassland;
NVC: W4, 17, 18; M1-3, 6, 8, 10, 11, 13-19, 21, 24, 25, 29; H1-8, 10-13, 16; CG11; U3, 5; MC10;
Ell.:

L	**F**	**R**	N	S
8	**8**	**2**	1	0

Gri.: S/SC

Erigeron acer (blue fleabane)

A species of open, well-drained, neutral or calcareous skeletal soils, brownfield sites and rock outcrops (especially chalk and limestone). Associated with coal waste.

Indicator: Often on a coal-related site (including railways, coal waste and smelting sites)

References: Kay (1994), Miller *et al.* (2007), Preston *et al.* (2002)

Ph. 1: B3 Calcareous grassland – unimproved, H6.5 Dune slack, I2 Artificial exposures and waste;
NVC: CG6, 7; SD16;
Ell.:

L	F	R	**N**	S
8	5	7	**3**	0

Gri.: SR/CSR

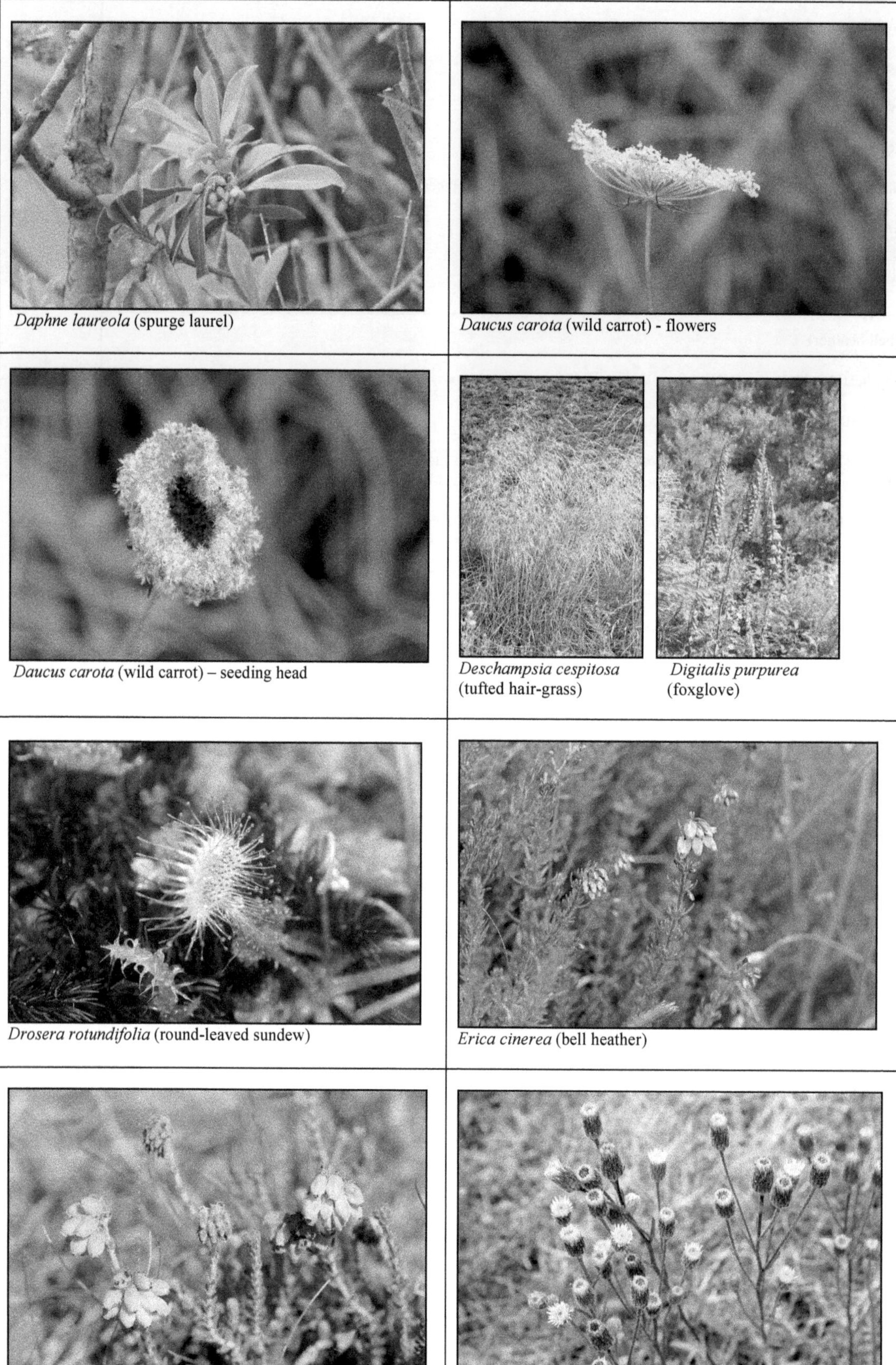

Daphne laureola (spurge laurel)

Daucus carota (wild carrot) - flowers

Daucus carota (wild carrot) – seeding head

Deschampsia cespitosa (tufted hair-grass)

Digitalis purpurea (foxglove)

Drosera rotundifolia (round-leaved sundew)

Erica cinerea (bell heather)

Erica tetralix (cross-leaved heath)

Erigeron acer (blue fleabane)

Euonymus europaeus (spindle)

A characteristic chalk shrub along with *Cornus sanguinea, Ligustrum vulgare* and *Viburnum lantana*. Variously described as a species of mostly calcareous habitats or those which are free-draining and base-rich.

Indicates: Calcareous or base rich soil

References: Clapham *et al.* (1962), Lousley (1969), Preston *et al.* (2002)

Ph. 1: A Woodland and scrub, J2 Boundaries;
NVC: W8, 12, 13, 21
Ell.:

L	F	**R**	N	S
5	5	**8**	5	0

Gri.: SC

Euphorbia amygdaloides (wood spurge)

A regional indicator of ancient semi-natural woodland (e.g. East Anglia) but sometimes also found in secondary woodland.

Indicator: A regional ancient woodland indicator (care needed)

References: Marren (1990)

Ph. 1: A1 Woodland (broad-leaved, semi-natural);
NVC: W8, 10, 12, 14;
Ell.:

L	F	R	N	S
4	5	6	6	0

Gri.: N/L

Fumaria species (fumitory) - images show *F. densiflora* and *F. occidentalis*

Weedy annuals. Seeds remain dormant until soil disturbance occurs. Most are fast colonizers of recently worked agricultural land, soon suppressed by later arrivals. The seeds probably remain viable in the soil for long periods.

Indicates: Recent soil disturbance

References: Smith (1984), Smith & Grenfell (1984)

Ph. 1: J1.1 Arable; C3.1 Tall ruderal; F2.2 Inundation (e.g. damp tracks)
NVC: OV4, 6, 9-11, 13-16, 18, 24, 33;
Ell.:

L	F	R	N	S
Varies with spp.				

Gri. (*F. officinalis*): R

Galium odoratum (woodruff)

Included on many English and Welsh regional lists of plants considered characteristic of ancient woodland (cf. *Melica uniflora*).

Indicator: Strong indicator of ancient semi-natural woodland

References: Rackham (2006)

Ph. 1: A Woodland and scrub;
NVC: W8-10, 12, 14;
Ell.:

L	F	R	N	S
3	5	7	6	0

Gri.: SC/CSR

Galium verum (lady's bedstraw)

A conspicuous species of unimproved neutral grassland, considered to be a 'less-exacting calcicole'.

Indicates: Unimproved neutral grassland (but also occurs in other swards)

References: Steele (1955)

Ph.1: A Woodland and scrub, B1 Acid grassland, B2 Neutral grassland, B3 Calcareous grassland, B5 Marsh/marshy grassland, C1 Bracken, D1 Dry dwarf shrub heath, F2.2 Inundation, H6.4 Dune slack, H6.5 Dune grassland, H6.6 Dune heath, H6.7 Dune scrub, H6.8 Open dune, H8.3 Crevice/ledge vegetation, H8.4 Coastal grassland;
NVC: W14, 19; M24; H6-8, 11; MG1, 2, 4, 5, 9; CG1-10, 13; U1, 4, 20; SD6-13, 16-19; MC5, 9, 11, 12; OV34;
Ell.:

L	F	**R**	**N**	S
7	4	**6**	**2**	0

Gri.: SC/CSR

Geum urbanum (wood avens)

Absent from places that dry out and from waterlogged ground.

Indicates: Moist sites

References:Taylor (1997)

Ph. 1: A1 Woodland and scrub;
NVC: W6-9, 12, 14, 21, 22, 24;
Ell.:

L	**F**	R	N	S
4	**6**	7	7	0

Gri.: CR/CSR

Glechoma hederacea (ground ivy)

Found in fertile habitats where other species are limited by shade and/or disturbance. May grow in open grassland or woodland.

Indicates: Shade or disturbance

References: Grime *et al.* (2007), Hutchings & Price (1999)

Ph. 1: A Woodland and scrub, B1.1 Acid grassland – unimproved, B2 Neutral grassland, C3.1 Tall ruderal, F2.1 Marginal, J2.5 Wall;
NVC: W2, 6-8, 10, 12, 13, 21, 24, 25; MG1; U1; S5, 26; SD15, 18; OV24, 25, 27;
Ell.:

L	F	R	N	S
6	6	7	7	0

Gri.: CR/CSR

Hedera helix (ivy)

A nitrophile, thriving in high nitrogen environments.

Indicates: High soil nitrogen

References: Pitcairn *et al.* (2006)
Ph. 1: Woodland and scrub, B2 Neutral grassland, C3.1 Tall ruderal, H8.4 Coastal grassland, J2 Boundaries (including J2.5 Wall);
NVC: W1-10, 12-17, 21-25; MG1; MC12; OV24, 27, 41, 42;
Ell.:

L	F	R	**N**	S
4	5	7	**6**	0

Gri.: SC

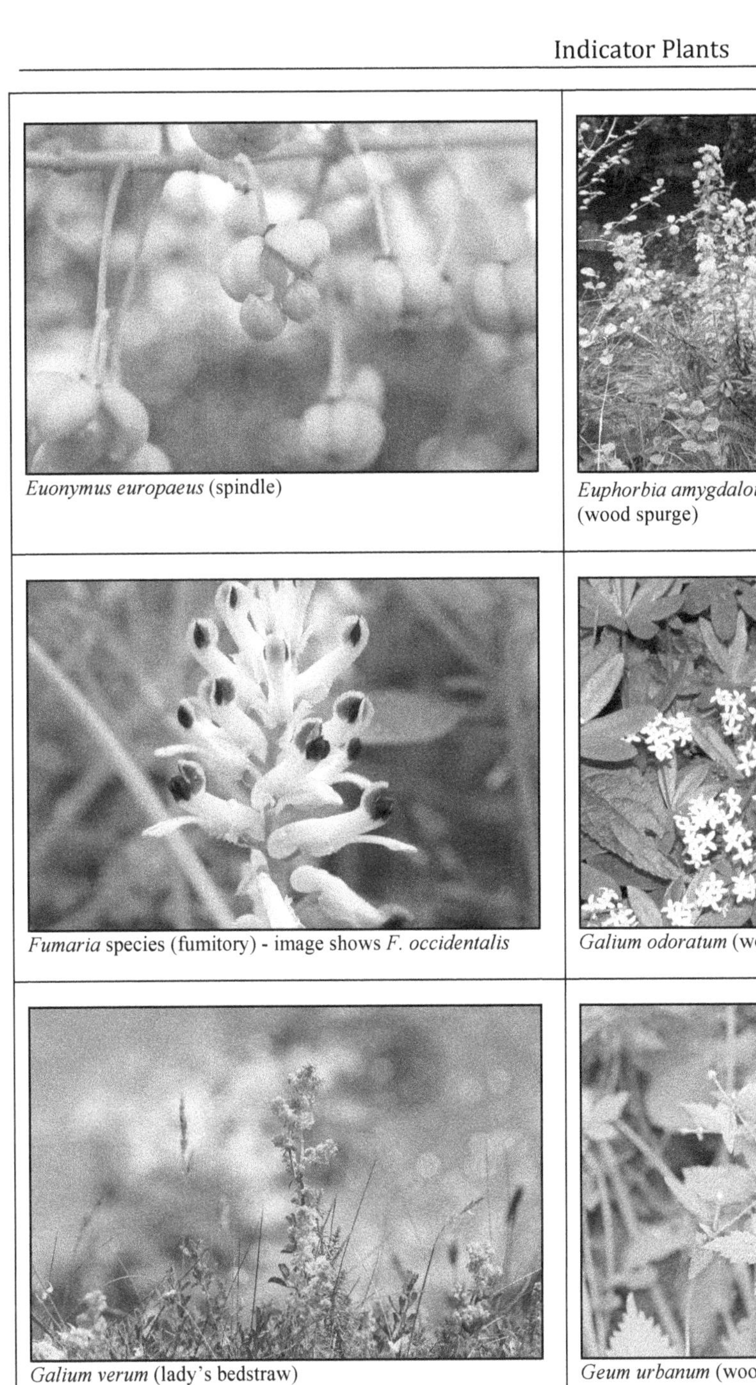

Euonymus europaeus (spindle)

Euphorbia amygdaloides (wood spurge)

Fumaria species (fumitory) - image shows *F. densiflora*

Fumaria species (fumitory) - image shows *F. occidentalis*

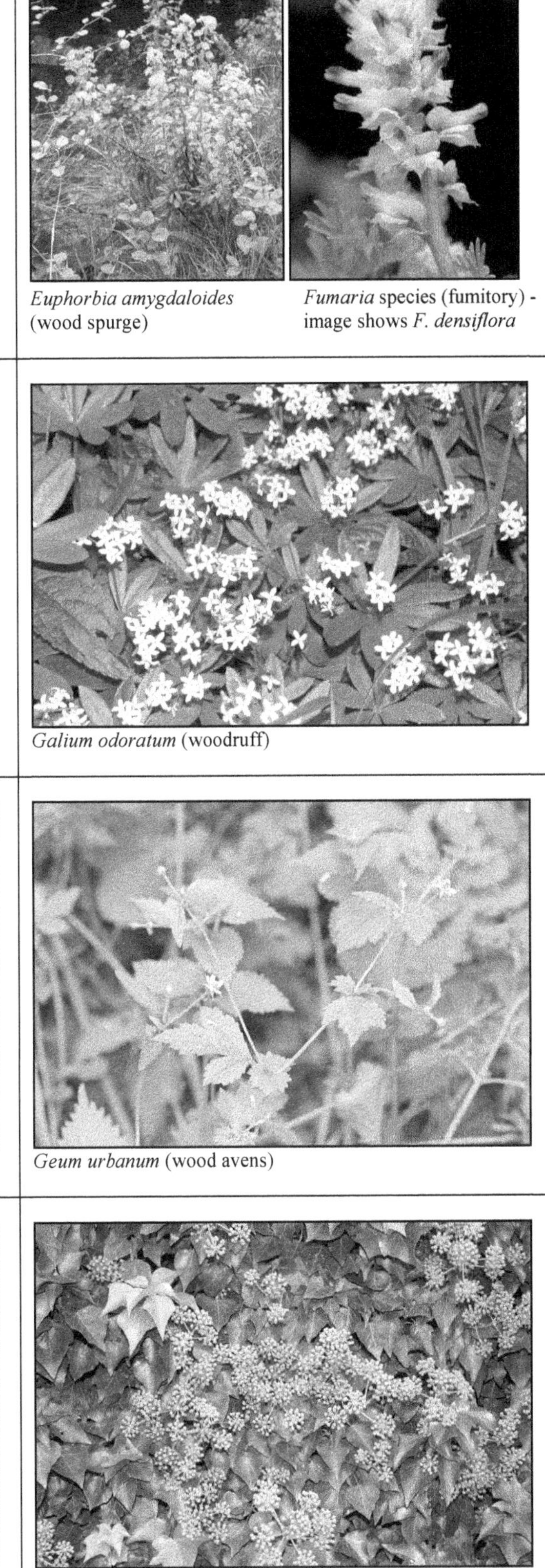

Galium odoratum (woodruff)

Galium verum (lady's bedstraw)

Geum urbanum (wood avens)

Glechoma hederacea (ground ivy)

Hedera helix (ivy)

Helianthemum nummularium (common rock-rose)

A nitrophobe.

Indicates: Low soil nitrogen

References: Pitcairn *et al.* (2006)
Ph. 1: B2 Neutral grassland, B3 Calcareous grassland, D1 Dry dwarf shrub heath, D5 Dry heath/acid grassland, H8.4 Coastal grassland, J2.5 Wall;
NVC: H4, 8; MG1; CG1-10; MC9, 11; OV39;
Ell.:

L	F	R	N	S
7	4	7	**2**	0

Gri.: S

Heracleum sphondylium (hogweed)

Present in many habitats (including woodland) but typical of infrequently grazed, neutral grassland.

Indicates: Infrequently grazed, neutral grassland

References: Sheppard (1991), Tansley (1949)

Ph. 1: A Woodland and scrub, B2 Neutral grassland, B3 Calcareous grassland – unimproved, C2 Upland species-rich ledges, C3.1 Tall ruderal, E3.2 Basin mire, F2.1 Marginal, H2 Saltmarsh, H6.5 Dune grassland, H6.7 Dune scrub, H8.4 Coastal grassland, J1 Cultivated/disturbed land;
NVC: W6, 8-10, 12, 21, 24, 25; M27; MG1-5, 9; CG14; U17; S23, 26; SM28; SD6-9, 18; MC9, 12; OV3, 6, 10, 12, 15, 18, 19, 21-27;
Ell.:

L	F	R	N	S
7	5	7	7	0

Gri.: C/CSR

Hordeum murinum (wall barley)

The seeds are carried on animals' fur and humans' clothes. The species is often plentiful at sites where dogs are exercised. High soil pH, high soil fertility, low competition and high light intensity benefit the species.

Indicates: A place where dogs are exercised

References: Davison (1971)

Ph. 1: B4 Improved grassland, C3.1 Tall ruderal, J1 Cultivated/disturbed land;
NVC: MG7; OV18, 22-25;
Ell.:

L	F	R	N	S
8	4	7	6	0

Gri.: R

Hyacinthoides non-scripta (bluebell)

A regional ancient woodland indicator (e.g. East England, Lincolnshire, Norfolk, Warwickshire). The large seeds are slow to reach new habitats.

Indicator: Regional ancient woodland indicator (especially East England)

References: Blackman & Rutter (1954), Knight (1964), Marren (1990), Rackham (2006)

Ph. 1: A1 Woodland and scrub, C1 Bracken, C2 Upland species-rich ledges, C3.1 Tall ruderal, H8.4 Coastal grassland;
NVC: W6-12, W14-17, 21, 22, 25; U17, 20; MC12; OV27
Ell.:

L	F	R	N	S
5	5	5	6	0

Gri.: SR

Hypericum calycinum (rose of Sharon)

A conspicuous remnant of urban planting (introduced species).

Indicates: Urban planting or escape from such

References: Crawley (2005)

Ph. 1: J1.4 Introduced shrub;
NVC: N/L;
Ell.:

L	F	R	N	S
5	7	5	5	0

Gri.: N/L

Jasione Montana (sheep's-bit)

A calcifuge of infertile sites.

Indicates: Acid infertile soils

References: Stace (2010), Steele (1955)

Ph. 1: A2 Scrub, D1 Dry dwarf shrub heath, H4 Rocks/boulders above high-tide mark, H6.6 dune heath, H8.3 Crevice/ledge vegetation;
NVC: W23; H6-8, 11; MC5;
Ell.:

L	F	R	N	S
7	4	4	**2**	0

Gri.: SR/CSR

Juncus species (rushes)

Wet and water-logged soils.

Indicates: Impeded drainage

Ph. 1: A Woodland and scrub, B1 Acid grassland, B2 Neutral grassland, B3.1 Calcareous grassland – unimproved, B5 Marsh/marshy grassland C3.2 Tall herb and fern – non-ruderal, D Heathland, E Mire, F Swamp, marginal and inundation, G1 Standing water, G2 Running water, H2 Saltmarsh, H5 Strandline vegetation, H6.4 Dune slack, H8.4 Coastal grassland, J1 Cultivated/disturbed land;
NVC: W1-8, 10, 21; M1, 4-32, 34, 35, 37, 38; H4, 5, 9, 10, 12-14, 16-20, 22; CG10-12; MG4-13; U2, 4-12, 14-16, 18, 19; S1, 2, 4-9, 11-15, 17-28; A7-11, 13, 14, 22-24; SD3; SD13-17; MC10; OV2, 6, 9, 19-21, 26-32, 34-36; SM13-20, 23, 24, 28;
Ell.:

L	F	R	N	S
Varies with spp.				

Gri.: Varies with species

Lamiastrum galeobdolon (yellow archangel)

An ancient woodland indicator species (not the variegated form).

Indicates: Ancient semi-natural woodland

Reference: Rackham (2006)

Ph. 1: A Woodland and scrub;
NVC: W7-10, 12, 14;
Ell.:

L	F	R	N	S
4	5	7	6	0

Gri.: S/SC

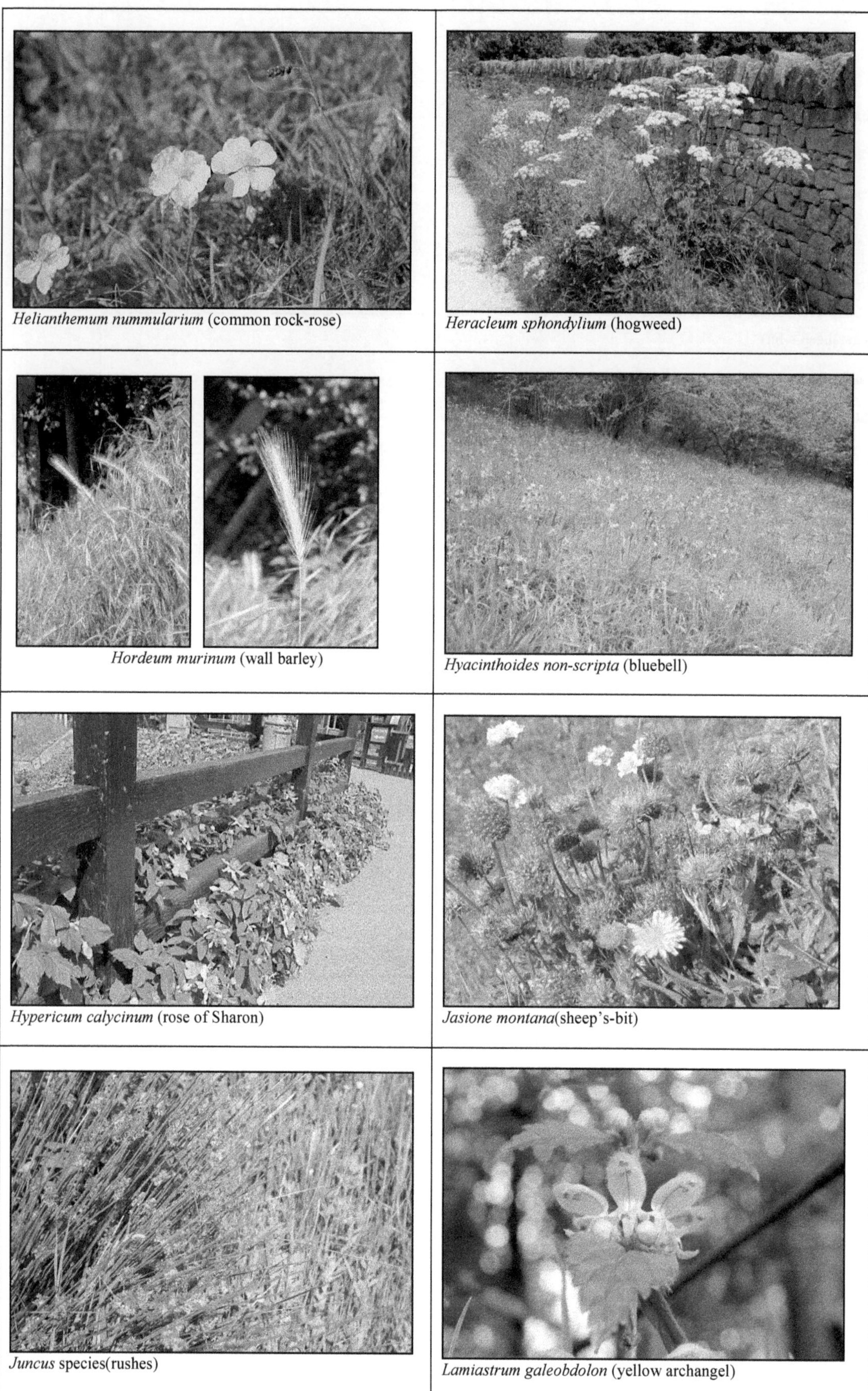

Helianthemum nummularium (common rock-rose)

Heracleum sphondylium (hogweed)

Hordeum murinum (wall barley)

Hyacinthoides non-scripta (bluebell)

Hypericum calycinum (rose of Sharon)

Jasione montana(sheep's-bit)

Juncus species(rushes)

Lamiastrum galeobdolon (yellow archangel)

Lathraea clandestina (purple toothwort)

Non-native, achlorophyllous, parasite. Occasionally cultivated in damp areas in large gardens. Seeds wash downstream to new hosts (willows (*Salix* spp.) and poplars (*Populus* spp.)). 10 years from seed to first flowers.

Indicator: In, or downstream from, a large garden (possibly distributed away from water by animals). In the wild it indicates habitat stability.

References: Atkinson (1996), Heinricher (1894), Lancaster (1990), Salisbury (1961)

Ph. 1: Usually by G2 Running water;
NVC:
Ell.: N/L;
Gri.: N/L

Lathraea squamaria (toothwort)

Native, achlorophyllous parasite. Hosts include elm (*Ulmus* spp.) and hazel (*Corylus avellana*). Probably confined to very old woodland and hedgerows. Development extremely slow, make take 10 years to flower.

Indicates: An old hazel coppice or woodland. Probably ancient woodland.

References: Kuijt (1969), Peterken (1974)

Ph. 1: A Woodland and scrub;
NVC: N/L;
Ell.:

L	F	R	N	S
3	6	7	6	0

Gri.: No strategy assigned

Leycesteria formosa (Himalayan honeysuckle)

A garden escape, sometimes planted for pheasant cover and naturalised.

Indicator: In woodlands, possibly an indicator of pheasant farming

References: Clement & Foster (1994)

Ph. 1: J1.4 Introduced shrub;
NVC: N/L;
Ell.:

L	F	R	N	S
6	5	7	6	0

Gri.: N/L

Ligustrum vulgare (wild privet)

Presence in quantity with dogwood (*Cornus sanguinea*) and wayfaring tree (*Viburnum lantana*) indicates a calcareous soil.

Indicates: Calcareous soil

References: Lousley (1969)

Ph. 1: A Woodland and scrub; J2 Boundaries;
NVC: W5, 8, 10, 12-14, 21;
Ell.:

L	F	**R**	N	S
6	5	**7**	5	0

Gri.: SC

Linaria vulgaris (common toadflax)

Base rich soil and disturbed sites.

Indicates: Disturbed, base-rich soil

Ph. 1: B3.1 Calcareous grassland – unimproved, H6.8 Open dune, J1.1 Arable;
NVC: CG6; SD6; OV16, 17;
Ell.:

L	F	**R**	N	S
7	4	**8**	6	0

Gri.: CR/CS

Linum catharticum (fairy flax)

A plant of dry, infertile calcareous/base-rich substrates but also flushed sites on neutral or mildly acidic soil. Associated with coal waste.

Indicates: Nutrient-poor open sites, often on brownfield substrates

References: Kay (1994), Miller *et al.* (2007), Preston *et al.* (2002), Stana & Vârban (2005)

Ph. 1: B2 Neutral grassland, B3.1 Calcareous grassland – unimproved, B5 Marshy grassland, C2 Upland species-rich ledges, D1 Dry dwarf shrub heath, D2 Wet dwarf shrub heath, D4 Montane heath/dwarf herb, E2.2 Basic flush, E2.3 Bryophyte dominated spring, E3.2 Basin mire, H6.4 Dune slack, H6.5 Dune grassland, H8.4 Coastal grassland;
NVC: M8, 10, 11, 13, 24, 38; H5, 8, 10; MG1; CG1-14; U15, 17; SD7, 8, 14, -17; MC10; OV37;
Ell.:

L	F	**R**	**N**	S
8	5	**7**	**2**	0

Gri.: SR

Luzula campestris (field wood-rush)

Presence in an old lawn indicates that reversion to a semi-natural grassland community has started and early colonizers such as daisy (*Bellis perennis*), dandelion (*Taraxacum officinale*) and annual meadow-grass (*Poa annua*) are declining.

Indicator: Presence in a lawn indicates that the sward has some antiquity

References: Gilbert & Anderson (1998)

Ph. 1: A1 Woodland, B1 Acid grassland, B2 Neutral grassland, B3 Calcareous grassland, B5 Marsh/marshy grassland, C1 Bracken, C3.1 Tall ruderal, D1 Dry dwarf shrub heath, D5 Dry heath/acid grassland, H6.4 Dune slack, H6.5 Dune grassland, H6.6 Dune heath, H8.4 Coastal grassland;
NVC: W11; M23, 25; H7, 10, 11, 16, 18; MG1, 3-6, 8; CG2, 4, 7, 10-14; U1, 3-5, 20; SD7-9, 11, 12, 16, 19; MC9, 10; OV27;
Ell.:

L	F	R	**N**	S
7	4	5	**2**	0

Gri.: S/CSR

Matricaria discoidea (pineapple-weed)

A species of the dusty, compacted soils beside paths and roads. Also spread in horse faeces and by mud on car wheels and shoes.

Indicates: Compacted soil

References: Lewandowski (2000), Salisbury (1961)

Ph. 1: J1 Cultivated/disturbed land;
NVC: N/L;
Ell.:

L	F	R	N	S
7	5	7	7	0

Gri.: R

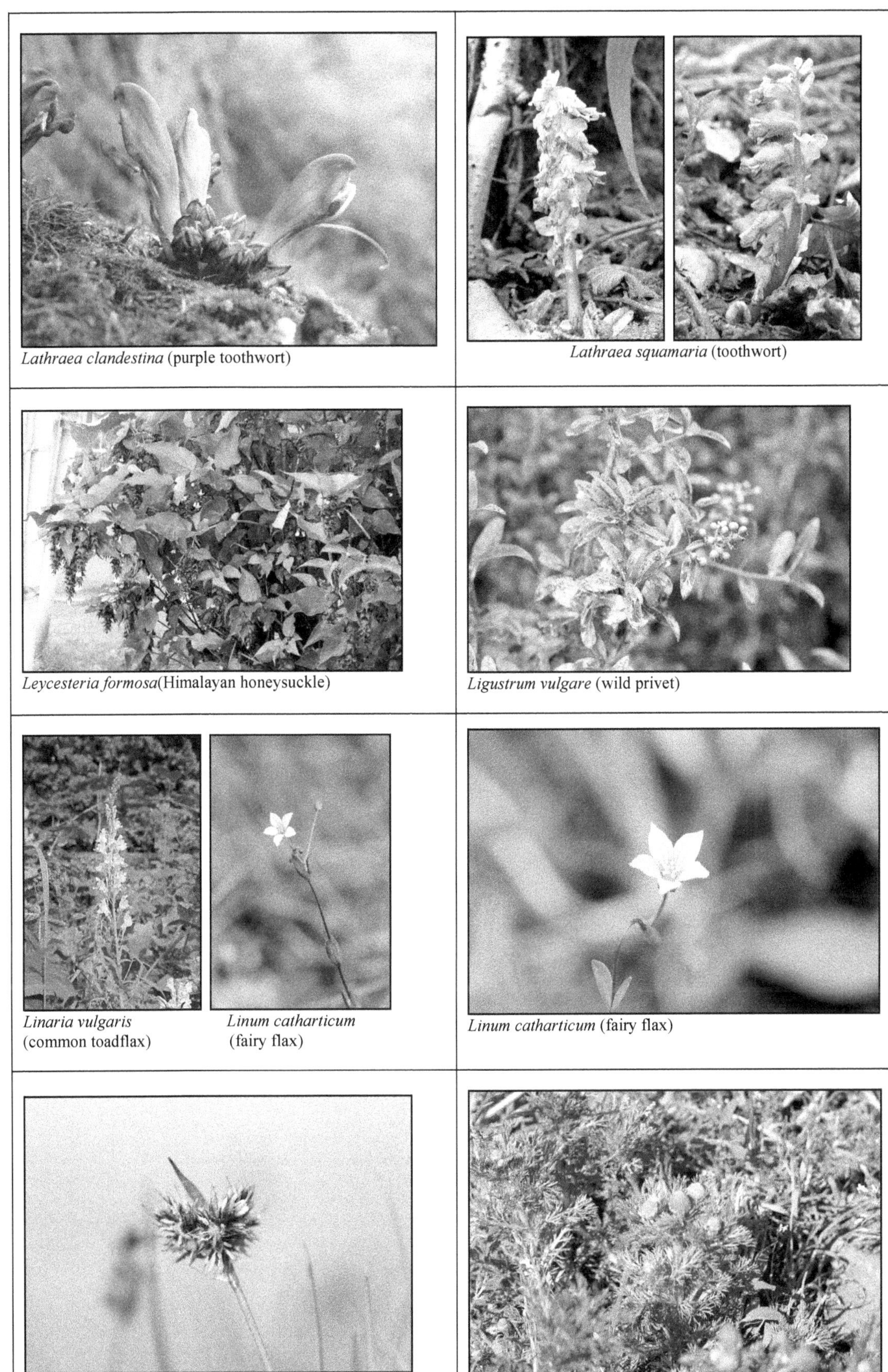

Lathraea clandestina (purple toothwort)

Lathraea squamaria (toothwort)

Leycesteria formosa(Himalayan honeysuckle)

Ligustrum vulgare (wild privet)

Linaria vulgaris (common toadflax)

Linum catharticum (fairy flax)

Linum catharticum (fairy flax)

Luzula campestris (field wood-rush)

Matricaria discoidea (pineapple-weed)

Melica uniflora (wood melick)

On many regional lists of plants characteristic of ancient woodland. All but one of the florets in each spikelet are adapted into an elaiosome and the species is dispersed by ants.

Indicator: Strong indicator of ancient woodland

References: Rackham (2006)

Ph. 1: A1 Woodland, J2 Boundaries;
NVC: W8-10, 12, 14;
Ell.:

L	F	R	N	S
4	5	7	5	0

Gri.: S/SC

Mercurialis perennis (dog's mercury)

Cannot grow on strongly acid or permanently waterlogged soils. On base rich soils it is often limited by poor drainage. Slow to colonize new woodland. Migration rates 0.2–0.6 m per yearby rhizome growth. Seed may be ejected to approximately 0.9 m away from the seed head, but a seedling takes 3 years to mature and spread by seed is slower than by rhizomes. Ants will carry seeds of *M. perennis* for longer distances.

Indicates: Older woodlands, avoiding strongly acid or waterlogged soils

References: Barnes & Williamson (2006), Brunet & von Oheimb (1998), Martin (1968), Mukerji (1936), Peterken & Game (1984), Pollard *et al.* (1974), Wilson (1968)

Ph. 1: A Woodland and scrub, B2 Neutral grassland, C2 Upland species-rich ledges, C3.1 Tall herb and fern – non-ruderal, C3.2 Tall herb and fern – non-ruderal, J2 Boundaries;
NVC: W6-10, 12-14, 19, 21, 22, 24, 25; MG1, 2; U17; OV27, 38;
Ell.:

L	F	R	N	S
3	6	7	7	0

Gri.: SC

Myosotis scorpioides (water forget-me-not)

Fertile, wetland sites. Replaced by *Myosotis secunda* where soils are infertile and base poor.

Indicates: Fertile, wet sites

References: Preston & Croft (1997)

Ph. 1: A Woodland and scrub, B2.2 Neutral grassland – semi-improved, B5 Marsh/marshy grassland, E3.2 Basin mire, F Swamp, marginal and inundation, G1.1 Standing water – eutrophic, G1.2 Running water – eutrophic;
NVC: W1, 3, 5, 6; M5, 22, 27; MG8, 10, 13; A5; S3-6, 11, 12, 14, 15, 17, 22-28; OV28, 29, 32, 35;
Ell.:

L	**F**	R	**N**	S
7	**9**	6	**6**	0

Gri.: CR

Neottia nidus-avis (bird's-nest orchid)

A mycoheterotrophic orchid, epiparasitic on trees via their own ectomycorrhizal fungi. Long development times from germination until production of first flower have been suggested. Included on some regional lists of indicators of ancient woodland in England and Wales.

Indicator: Complex life cycle argues for long-term habitat stability

References: Björkman (1960), Leake (2004), McKendrick *et al.* (2002), Rackham (2006), Selosse *et al.* (2002), Summerhayes (1968)

Ph. 1: A1.1.1 Broadleaved woodland – semi-natural, A2 Scrub;
NVC: W12;
Ell.:

L	F	R	N	S
2	4	7	5	0

Gri.: No strategy assigned

Noccaea caerulescens (Alpine pennycress)

An absolute metallophyte of metalliferous soils in the UK. Able to hyper-accumulate zinc, cadmium and nickel and thrive on metal contaminated soils (including lead and copper). This species, vernal sandwort ('leadwort') (*Minuartia verna*), Pyrenean scurvy-grass (*Cochlearia pyrenaica*), sea campion (*Silene uniflora*) and thrift (*Armeria maritima*) are notable metallophytes.

Indicator: Universal indicator of metal contamination

References:DoE (1994), Robinson *et al.* (1998), Sellars & Baker (1988), Simkin (2007)

Ph. 1: B3.1 Calcareous grassland – unimproved (Calaminarian grassland), I Rock exposure and waste;
NVC: OV37;
Ell.:

L	F	R	**N**	S
8	4	6	**1**	0

Gri.: N/L

Ophrys apifera (bee orchid)

Prefers dry, base-rich turf over calcareous substrates including anthropogenic habitats e.g. roadsides. Potentially 8 years from seed to flower. Populations can fluctuate significantly in numbers of individuals from year to year.

Indicates: Dry (usually), base-rich habitat, sometimes anthropogenic

References: Foley & Clarke (2005), Mabey (1996), Summerhayes (1968)

Ph. 1: B3 Calcareous grassland - unimproved;
NVC: CG2, 3, 5;
Ell.:

L	**F**	**R**	**N**	S
8	**4**	**8**	**3**	0

Gri.: SR

Orchis mascula (early purple orchid)

Sometimes considered an ancient woodland species. At least 4 years from first appearance above ground to first flowers.

Indicates: Ancient woodland indicator but also found in grassland. If in flower, the habitat has been suitable for at least 5 years. Habitat stability.

References: Hermy *et al.* (1999), Jacquemyn *et al.* (2009), Kretzschmar *et al.* (2007), Möller (1987)

Ph. 1: A1 Woodland, B2.2 Neutral grassland – unimproved, B3.1 Calcareous grassland – unimproved, C2 Upland species-rich ledges;
NVC: W8; MG3; CG13; U17;
Ell.:

L	F	R	N	S
6	5	7	4	0

Gri.: SR

Orobanche species (broomrapes)

Obligate parasites, little/no chlorophyll, host-dependent. Flower beds, crop fields, semi-natural habitats. New site records may indicate recent arrival of parasite or its host. Some are agricultural pests. Small seeds can remain dormant for over 10 years, waiting for a host plant.

Indicator: A prompt. Indicates a need to ask questions, the answers to which may reveal insights into habitat history. What is the host? How and when did the broomrape arrive? Consider any recent environmental changes.

References: Cubero & Moreno (1979), Linke *et al.*, (1989), Lopez-Granados & Garcia-Torres (1999), Salisbury (1961)

Ph. 1: Various, including B Grassland, D Heathland, J1 Cultivated/disturbed land, J2 Boundaries;
NVC: CG2 (*O. picridis*), 3, 6 (*O. elatior*);
Ell.:

L	F	R	N	S
Varies with species				

Gri.: N/L

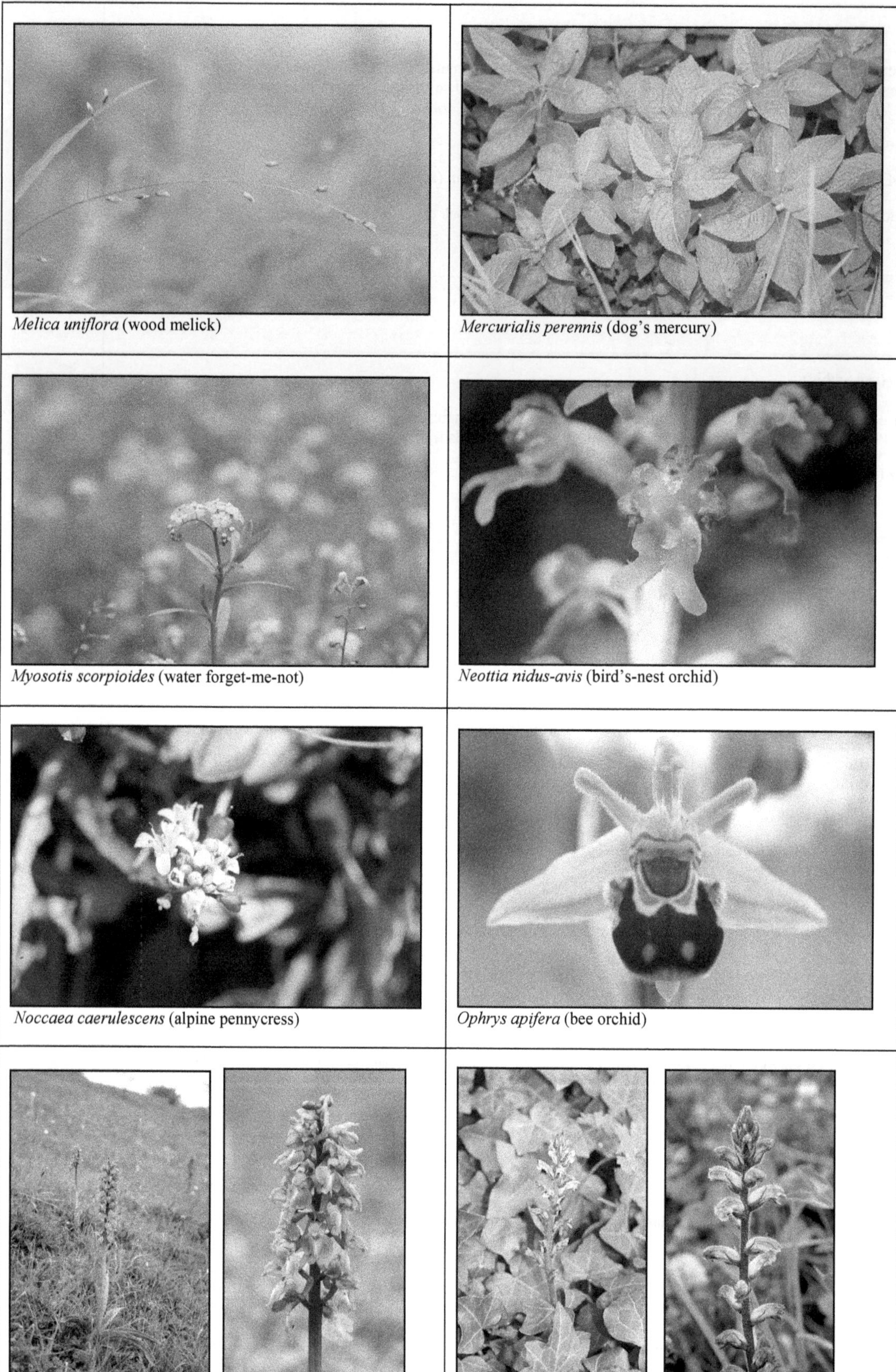

Melica uniflora (wood melick)

Mercurialis perennis (dog's mercury)

Myosotis scorpioides (water forget-me-not)

Neottia nidus-avis (bird's-nest orchid)

Noccaea caerulescens (alpine pennycress)

Ophrys apifera (bee orchid)

Orchis mascula (early purple orchid)

Orobanche hederae and *O. minor* (broomrapes)

Oxalis acetosella (wood sorrel)

An ancient woodland indicator species.

Indicates: Ancient semi-natural woodland

References: Rackham (2006)

Ph. 1: A Woodland and scrub, B1 Acid grassland, B2.1 Neutral grassland – unimproved, B3.1 Calcareous grassland – unimproved, C1 Bracken, C2 Upland species-rich ledges, C3.2 Tall herb and fern – non-ruderal, D1.1 Dry dwarf shrub heath – acid, D4 Montane heath/dwarf herb;
NVC: W4, 7-11, 14-20, 22; H18, 20, 21; MG2; CG10-12; U4, 13, 15, 16-21; OV38;
Ell.:

L	F	R	N	S
4	6	4	4	0

Gri.: S/SR

Paris quadrifolia (herb-paris)

Ancient woodland indicator (sometimes in younger habitats). Average rate of spread up to 0.33 m per year by seeds, also spreads by rhizomes. Absent from very dry and wet places.

Indicator: Strong indicator of ancient semi-natural woodland and moist sites

References: Beckett & Bull (1999), Brunet & von Oheimb (1998), Ellenberg (1988), Honnay *et al.* (1999), Jacquemyn *et al.* (2008), Peterken (1981), Salisbury (1942), Thompson *et al.* (1997)

Ph. 1: A1 Woodland;
NVC: N/L;
Ell.:

L	F	R	N	S
3	6	7	6	0

Gri.: SR/CSR

Plantago major (great plantain)

Thrives on human tracks. Known as English-man's Foot by native American Indians.

Indicator: Trample-resistant and therefore often on tracks and trampled places where other species cannot compete.

References: Grigson (1958), Josselyn (1672)

Ph. 1: B2 Neutral grassland, B3.1 Calcareous grassland – unimproved, B4 Improved grassland, B5 Marsh/marshy grassland, C1 Bracken, C3.1 Tall ruderal, F Swamp, marginal and inundation, H6.4 Dune slack, H6.5 Dune grassland, J1 Cultivated/disturbed land;
NVC: MG6, 7, 10-12; CG5; S18, 23; SD8, 14; OV4, 7-13, 15, 16, 18-24, 28-31, 33, 36;
Ell.:

L	F	R	**N**	S
7	5	6	**7**	0

Gri.: R/CSR

Potentilla erecta (tormentil)

A nitrophobe and acidity indicator.

Indicates: Acidic substrate and low soil nitrogen

References: Pitcairn *et al.* (2006)

Ph. 1: A Woodland and scrub, E Mire, B1.1 Acid grassland – unimproved, B2.1 Neutral grassland – unimproved, B3.1 Calcareous grassland – unimproved, C1 Bracken, C2 Upland species-rich ledges, C3.1 Tall ruderal, C3.2 Tall herb and fern – non-ruderal, D Heathland (including H6.6 Dune heath), D4 Montane heath/dwarf herb, E3.2 Basin mire, H6.5 Dune grassland, H8.4 Coastal grassland;
NVC: W2, 4, 9, 11, 16-19, 23, 25; M4, 6, 8-17, 19, 21-26, 29, 38; H2-22; MG2, 3, 5, 9; CG2, 9-14; U1-7, U10, 13-17, 19-21; S24; SD12; MC8-10, 12; OV27;
Ell.:

L	F	**R**	**N**	S
7	7	**3**	**2**	0

Gri.: S/CSR

Primula veris (cowslip)

Increased in recent decades in parts of UK, because its seeds are included in wildflower seed mixtures sown on motorway verges, embankments and urban conservation areas.

Indicates: Neutral or calcareous grassland but, in a transport corridor or urban site, it may indicate deliberate sowing of seed

References: Brys & Jacquemyn (2009)

Ph. 1: B2 Neutral grassland, B3 Calcareous grassland, H6.5 Dune grassland;
NVC: MG1, 3-5, 9; CG2-6, 8; SD8, 9;
Ell.:

L	F	R	N	S
7	4	7	3	0

Gri.: S/CSR

Prunus spinosa (blackthorn)

Specimens with unusually large fruits may be *P. domestica* ssp. *insititia* (bullace, damson) or ssp. *italica* (greengage). These might be indicative of an old orchard or planting exercise. Extensive hybridization has rendered separation of the subspecies difficult.

Indicates: Often listed *Prunus* sp. in ecological surveys, but large fruits may indicate former cultivation

References: Stace (2010)

Ph. 1: A Woodland and scrub, D1 Dry dwarf shrub heath, J2 Boundaries;
NVC: W6-8, 10, 21, 22, 24, 25; H6;
Ell.:

L	F	R	N	S
6	5	7	6	1

Gri.: SC

Pyracantha coccinea (firethorn)

Often used as a security screen due to its unpleasant thorns but may spread by seed to other sites. Blackbirds defend the bushes in winter for the rich crop of orange red berries.

Indicates: Former urban planting where the control of access is/was important (see also *Berberis darwinii*)

References: Crawley (2005)

Ph. 1: J1.4 Introduced shrub;
NVC: N/L;
Ell.: N/L;
Gri.: N/L

Ranunculus acris (meadow buttercup)

Characteristic of grazed/mown grasslands with a good supply of water. Abundance in pastures related to pasture age. Frequency increases with grazing as it is avoided by many herbivores. Survives in hay meadows as tall stems allow competition with tall grasses. Occupies sites where drainage conditions are intermediate between those occupied by *R. bulbosus* and *R. repens*. In ridge and furrow, *R. acris* occupies the sides, *R. bulbosus* the tops and *R. repens*, the furrows.

Indicates: Moist, grazed/cut grassland

References: Harper (1957)

Ph. 1: A Woodland and scrub, B1 Acid grassland, B2 Neutral grassland, B3.1 Calcareous grassland – unimproved, B5 Marsh/marshy grassland, C1 Bracken; C2 Upland species-rich ledges, C3.1 Tall ruderal, D4 Montane heath/dwarf herb, E2.2 Basic flush, E2.3 Bryophyte dominated spring, E3.2 Basin mire, H2 Saltmarsh, H6.4 Dune slack, H6.5 Dune grassland, H8.3 Crevice/ledge vegetation, H8.4 Coastal grassland;
NVC: W3, 5-9, 11, 19, 20, 24; M8-10, 12, 13, 22-28, 31, 32, 34, 38; H18; MG1-12; CG10-14; U4, 5, 13-15, 17, 20; S5, 7, 11, 19, 26; SM16, 18; SD8, 14-16; MC5, 9-11; OV26;
Ell.:

L	**F**	R	N	S
7	**6**	6	4	0

Gri.: CSR

Oxalis acetosella (wood sorrel)

Paris quadrifolia (herb-paris)

Plantago major (great plantain)

Potentilla erecta (tormentil)

Primula veris (cowslip)

Prunus spinosa (blackthorn)

Pyracantha coccinea (firethorn)

Ranunculus acris (meadow buttercup)

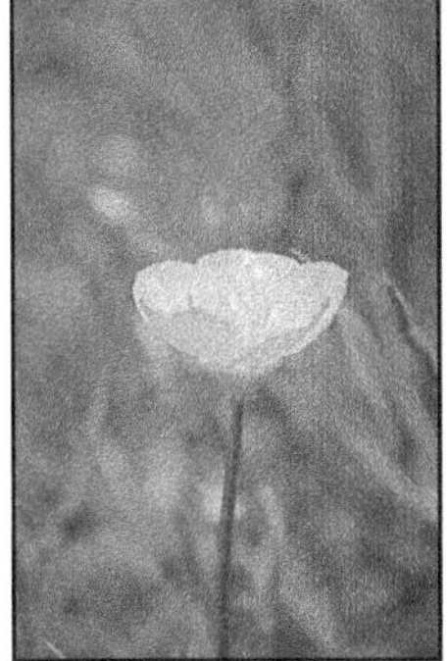
Ranunculus bulbosus (bulbous buttercup)

R. bulbosus (bulbous buttercup)

Characteristic of well-drained, *sunny*, grazed grasslands. Grazing encourages spread (it is avoided by cattle) but it does not persist in hay meadows where it is out-competed by tall grasses. Excluded by poor drainage and trampling.

Indicates: Well-drained, sunny, grazed grassland

References: Harper (1957)

Ph. 1: A1 Woodland, B2 Neutral grassland, B3 Calcareous grassland, B4 Improved grassland, H6.5 Dune grassland, H8.4 Coastal grassland, J1.2 Amenity grassland;
NVC: W12; MG1, 3-5, 7; CG1-7; SD7-9; MC11; OV23;
Ell.:

L	F	R	N	S
7	**4**	7	4	0

Gri.: SR

R. repens (creeping buttercup)

Species of heavy, wet soils, especially abundant where drainage is impeded and on disturbed soils. More palatable to stock than *R. acris* or *R. bulbosus*. Age of meadows has been related to the number of *R. repens* flowers present with additional petals (>5).

Indicates: Impeded drainage, disturbed soil

References: Harper (1957), Warren (2009)

Ph. 1: A Woodland and scrub, B1 Acid grassland, B2.1 Neutral grassland – unimproved, B2.2 Neutral grassland – semi-improved, B3 Calcareous grassland, B4 Improved grassland, B5 Marsh/marshy grassland, C1 Bracken, C3.1 Tall ruderal, E2.2 Basic flush, E2.3 Bryophyte dominated spring, F Swamp, marginal and inundation, H6.4 Dune slack, H6.5 Dune grassland, J1 Cultivated/disturbed land;
NVC: W1, 3, 5-11, 14, 21, 24; M22, 23, 27, 28, 35, 38; MG1, 3-11, 13; CG3, 4, 10; U4, 20; S5, 7, 12, 15, 16, 18, 19, 23, 26-28; SD6-8, 12, 14-17; OV2, 4-7, 10-12, 15, 19-23, 25, 26, 28-33, 36;
Ell.:

L	F	R	N	S
6	**7**	6	7	0

Gri.: CR

Rhinanthus minor (yellow-rattle)

A hemi-parasitic species of neutral and basic grasslands. Hemi-parasitic flowering plants characteristically occur in low-nutrient habitats.

Indicates: Low soil fertility sward (or a deliberate introduction to reduce vigour of coarse grasses)

References: ter Borg (1972), Davies & Graves (2000), Parker & Riches (1993)

Ph. 1: A2 Scrub, B2.1 Neutral grassland – unimproved, B2.2 Neutral grassland – semi-improved, B3 Calcareous grassland, B5 Marsh/marshy grassland, C2 Upland species-rich ledges, D4 Montane heath/dwarf herb, E3.2 Basin mire, H6.4 Dune slack, H6.5 Dune grassland;
NVC: W20; M22, 26; MG1, 3-6, 8; CG1-3, 5, 8, 14; U15, 17; S24; SD7, 8, 14, 16, 17;
Ell.:

L	F	R	N	S
7	5	6	**4**	0

Gri.: R/SR

Rubus fruticosus (bramble)

A nitrophile, which thrives in high nitrogen environments.

Indicates: High soil nitrogen

Reference: Pitcairn *et al.* (2006)

Ph. 1: A Woodland and scrub, B1 Acid grassland, B2 Neutral grassland, B5 Marsh/marshy grassland, C3.1 Tall ruderal, D1 Dry dwarf shrub heath, F1 Swamp, F2.1 Marginal, H6.5 Dune grassland, H6.7 Dune scrub, H6.8 Open dune, H8.4 Coastal grassland, J1.3 Ephemeral/short perennial;
NVC: W1-17, 21-25; M24, 25, 27; H2, 4, 6; MG1, 9; U1, 2; S3, 4, 6, 17, 18, 24-26, 28; SD6, 7, 18; MC12; OV22, 24-27;
Ell.:

L	F	R	N	S
6	6	6	**6**	0

Gri.: SC

Sambucus nigra (*elder*)

Characteristic of disturbed, base-rich and nitrogen-rich soils and locally enriched areas around rabbit warrens/badger setts. Possibly a phosphate plant. The NVC sycamore–elder subcommunity of the rosebay willowherb community (OV27) can indicate where fires have occurred in woodland. It is resistant to some forms of pollution and used as a street tree in Ecuador.

Indicates: Eutrophicated, disturbed soils

References: Atkinson & Atkinson (2002), Clapham *et al.* (1962), Rackham (1986), Rodwell (2000)

Ph. 1: A Woodland and scrub, C3 Calcareous grassland, C3.1 Tall ruderal, H6.7 Dune scrub, J2 Boundaries;
NVC: W6-10, 12-14, 21, 24, 25; (CG4, 5, 7; SD18; OV27 as saplings);
Ell.:

L	F	R	N	S
6	5	7	7	0

Gri.: C

Sanguisorba minor (salad burnet)

Characteristic species of calcareous grassland (but sometimes found in neutral grassland).

Indicates: Predominately indicates calcareous conditions

References: Lousley (1969), Stace (2010), Zohlen & Tyler (2004)

Ph. 1: A2 Scrub, B2.1 Neutral grassland – unimproved, B2.2 Neutral grassland – semi-improved, B3 Calcareous grassland, D1 Dry dwarf shrub heath – basic (limestone heath), H8.5 Coastal heathland, B1 Acid grassland (calcareous influence), H6.5 Dune grassland, H8.3 Crevice/ledge vegetation, H8.4 Coastal grassland, I Rock exposure and waste (basic), J2.5 Wall;
NVC: W21; H8; MG1, 5; CG1-9; U4; SD9; MC5, 11; OV38-41;
Ell.:

L	F	**R**	N	S
7	4	**8**	3	0

Gri.: S/CSR

Sanicula europaea (sanicle)

Associated with ancient woodland in southern England. Grows readily from seed and the hooked fruits are dispersed on clothes so perhaps this is a species which has either lost its principal vector or suffered a limitation to its dispersal as a result of habitat fragmentation.

Indicator: Regional ancient woodland indicator

References: Marren (1990) citing Hornby & Rose (1986)

Ph. 1: A1 Woodland (broadleaved, semi-natural);
NVC: W8-10, 12;
Ell.:

L	F	R	N	S
4	5	7	5	0

Gri.: S/CSR

Saxifraga tridactylites (rue-leaved saxifrage)

Typical of dry soils.

Indicates: Dry soil (drought-stressed environment)

Ph. 1: I Rock exposure and waste (basic); J2.5 Wall;
NVC: OV38, 39, 42;
Ell.:

L	**F**	R	**N**	S
7	**2**	7	**2**	0

Gri.: SR

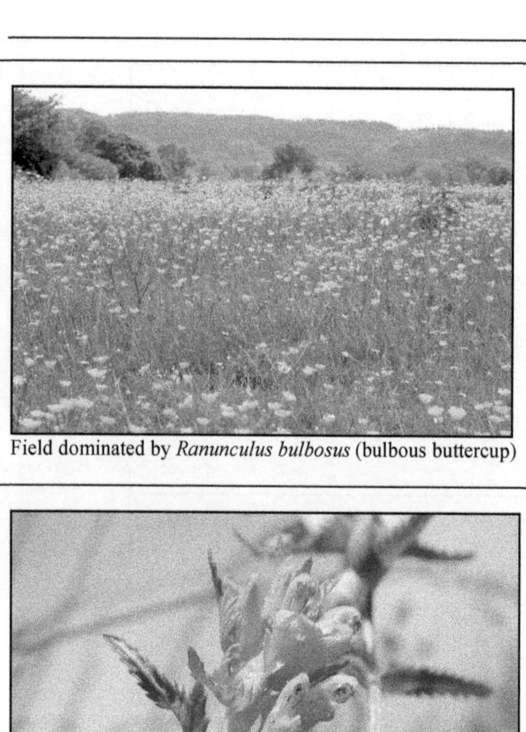
Field dominated by *Ranunculus bulbosus* (bulbous buttercup)

Ranunculus repens (creeping buttercup)

Rhinanthus minor (yellow-rattle)

Rubus fruticosus (bramble)

Sambucus nigra (elder)

Sanguisorba minor (salad burnet)

Sanicula europaea (sanicle)

Sanicula europaea (sanicle)

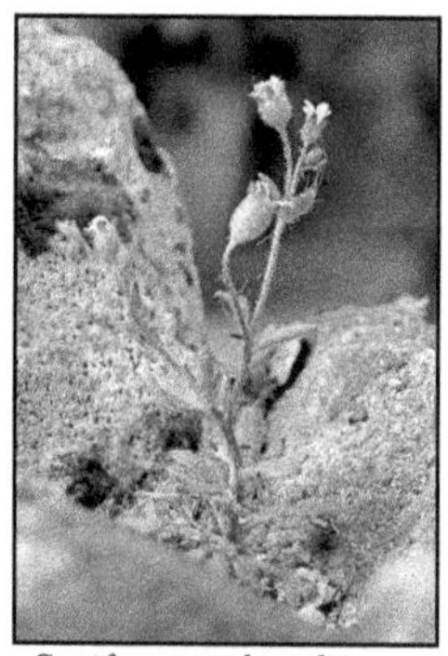
Saxifraga tridactylites (rue-leaved saxifrage)

Senecio jacobaea (ragwort)

Particularly common on poor light soils and often a feature in the development of sand dunes. Selective feeding by rabbits (which avoid ragwort) may benefit its spread. *Hyoscyamus niger* (henbane) and hemlock (*Conium maculatum*) are immune to rabbit browsing and may occur around their burrows. Poisonous to livestock and controlled under the Weeds Act 1959.

Indicates: Light soils

References: Defra (2004), Salisbury (1961)

Ph. 1: A Woodland and scrub, B1 Acid grassland, B2 Neutral grassland, B3 Calcareous grassland, D5 Dry heath/acid grassland, F Swamp, marginal and inundation, H3 Shingle/gravel above high-tide mark, H6.4 Dune slack, H6.5 Dune grassland, H6.6 Dune heath, H6.7 Dune scrub, H6.8 Open dune;
NVC: W9, 10, 23, 24; H1, 11; MG1-3, 5, 9; CG2-8, 10, 13; U1, S23; SD1, 4, 6-16, 18, 19; MC4, 5, 9, 11; OV10, 19-21, 23, 25, 27, 37, 38;
Ell.:

L	F	R	N	S
7	4	6	4	0

Gri.: CR/CSR

Silene uniflora (sea campion)

Largely a coastal plant, its inland distribution is confined to base-rich, mountain rocks or metalliferous mine sites.

Indicates: When inland, consider heavy metal contamination

References: Baker & Dalby (1980)

Ph. 1: A2 Scrub, D1.2 Dry dwarf shrub heath – basic, F2.2 Inundation, H2 Saltmarsh, H3 Shingle/gravel above high-tide mark, H5 Strandline vegetation, H8.3 Crevice/ledge vegetation, H8.4 Coastal grassland, I2 Spoil heap;
NVC: W22; H7; MG11; SM28; SD1, 2; MC1-6, 8-12;
Ell.:

L	F	R	N	S
8	6	6	4	3

Gri.: N/L

Sisymbrium officinale (hedge mustard)

Associated with human habitation and anthropogenic disturbance. Possibly a useful indicator for human disturbance in woodlands.

Indicates: Human disturbance

References: Rich (1991), Tsuyuzaki & Titus (2010)

Ph. 1: C3.1 Tall ruderal, J1 Cultivated/disturbed land;
NVC: OV3, 7-9, 13, 16-19, 22-25;
Ell.:

L	F	R	N	S
7	4	7	7	0

Gri.: R/CR

Smyrnium olusatrum (alexanders)

Cultivated as a vegetable until replaced by celery in the fifteenth century. Its predominantly coastal distribution is puzzling as some inland populations have persisted for decades.

Indicates: If inland, consider possible former site of cultivation

References: Preston *et al.* (2002), Salisbury (1961)

Ph. 1: C3.1 Tall ruderal;
NVC: OV24;
Ell.:

L	F	R	N	S
7	5	7	7	0

Gri.: N/L

Solanum nigrum (black nightshade)

Typical of moist sites

Indicates: A moist soil

Reference: Hill *et al.* (2004)

Ph. 1: F1 Swamp, F2.1 Marginal, J1.1 Arable;
NVC: S8; OV4, 5, 7-9, 11, 13, 14, 17, 19;
Ell.:

L	F	R	N	S
7	**5**	7	8	0

Gri.: R/CR

Sparganium erectum (branched bur-reed)

Tolerant of eutrophication but palatable to stock and often absent from wetland edges to which cattle have access, while frequent on adjacent, ungrazed stretches.

Indicates: Ungrazed wetland margins

References: Preston & Croft (1997)

Ph. 1: A1 Woodland, C3.1 Tall ruderal, F Swamp, marginal and inundation, G1.1 Standing water – eutrophic, G1.2 Standing water – mesotrophic, G1.3 Standing water – oligotrophic, G1.4 Standing water – dystrophic, G2.2 Running water – mesotrophic, G2.3 Running water – oligotrophic, G2.4 Running water – dystrophic;
NVC: W5; M27; A4, 9, 20; S3-8, 12, 14-18, 22, 26, 27; OV26, 30;
Ell.:

L	F	R	N	S
7	**10**	7	7	0

Gri.: C/CR

Symphoricarpos species (snowberry)

A member of the urban shrub planting assemblage common on industrial estates, schools and other town and city planting schemes

Indicates: Former urban planting, escape or discard

References: Crawley (2005)

Ph. 1: J1.4 Introduced shrub;
NVC: N/L;
Ell. (*S. albus*):

L	F	R	N	S
5	5	6	7	0

Gri. (*S. albus*): C/SC

Trifolium repens (white clover)

Shade intolerant and nutrient-demanding. Occurs in a range of habitat types but uncommon in species-rich, old chalk grassland. When present in such grassland, it probably indicates enrichment with fertilizer or faeces.

Indicates: When in species-rich, old chalk grassland, a marker for areas of nutrient enrichment

References: Burdon (1983), Wells (1975)

Ph. 1: A2 Scrub, B1 Acid grassland, B2 Neutral grassland, B3 Calcareous grassland, B5 Marsh/marshy grassland, C1 Bracken, D4 Montane heath/dwarf herb, E2.2 Basic flush, E2.3 Bryophyte dominated spring, F1 Swamp, H2 Saltmarsh, H6.4 Dune slack, H6.5 Dune grassland, H6.6 Dune heath, H6.8 Open dune, H8.3 Crevice/ledge vegetation, H8.4 Coastal grassland, H8.5 Coastal heathland, J1.1 Arable;
NVC: W24, M13, 22, 23, 26, 35, 38; H7, 11; MG1, 3-12; CG1-4, 6-11, 13; U1, 4-6, 15, 20; S19; SM16, 18-20, 28; SD6-9, 12, 14-17, 19; MC5, 8-11; OV1-6, 8-13, 15, 16, 19-23, 25, 28, 30, 32, 33, 37;
Ell.:

L	F	R	N	S
7	5	6	6	0

Gri.: CR/CSR

 *Senecio jacobaea* (ragwort)	 *Silene uniflora* (sea campion)
 *Sisymbrium officinale* (hedge mustard)	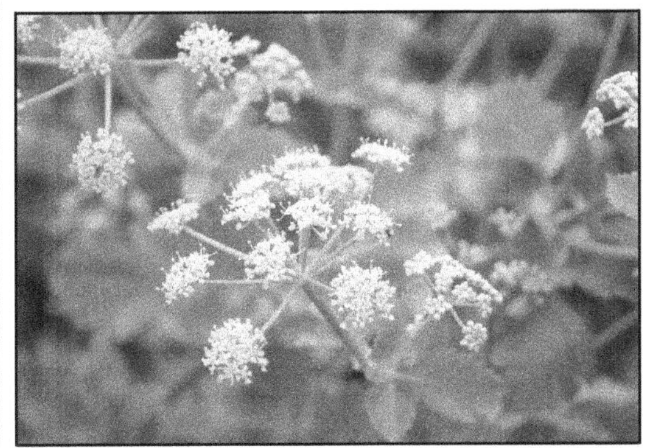 *Smyrnium olusatrum* (alexanders)
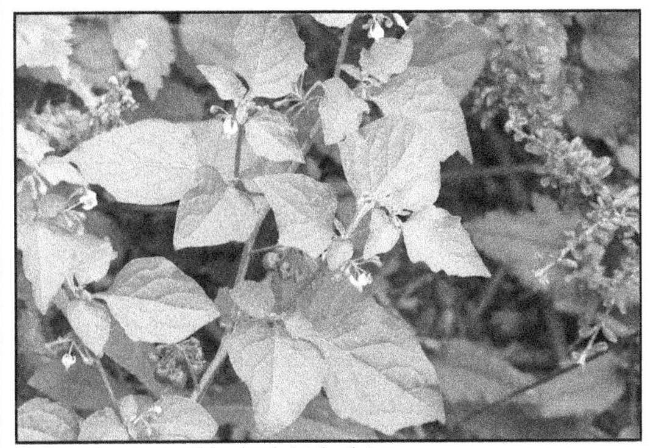 *Solanum nigrum* (black nightshade)	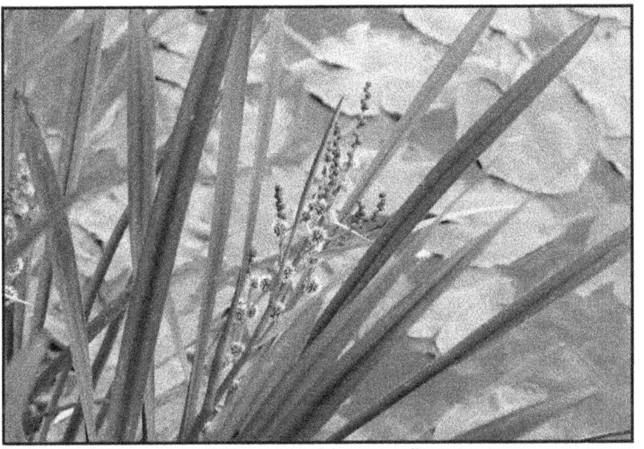 *Sparganium erectum* (branched bur-reed)
 *Symphoricarpos* species (snowberry)	 *Trifolium repens* (white clover) 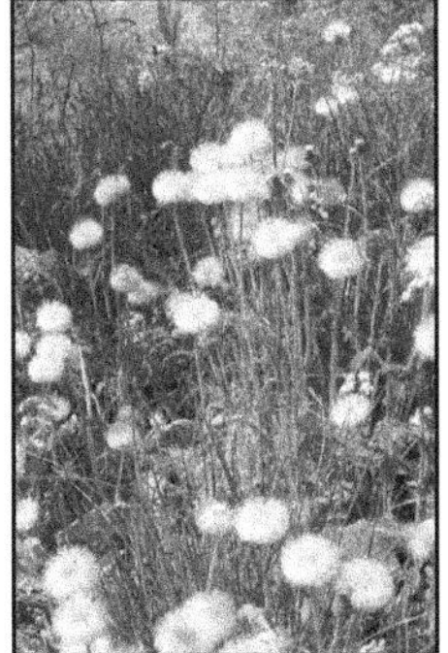 *Tussilago farfara* (coltsfoot, claywort)

Tussilago farfara (coltsfoot, claywort)

A rhizomatous weed of anthropogenic sites including quarries, road verges and other brownfield habitats. It will grow in a variety of soils, especially if moist, but thrives in heavy, calcareous clay.

Indicates: Heavy, damp soils

References: Salisbury (1961)

Ph. 1: A Woodland and scrub, C2 Upland species-rich ledges, D4 Montane heath/dwarf herb, C3.1 Tall ruderal, C3.2 Tall herb and fern – non-ruderal, E2.3 Bryophyte dominated spring, H6.5 Dune grassland, H6.8 Open dune, I2 Artificial exposures and waste, J1 Cultivated/disturbed land;
NVC: W7, 9; M11, 37; U15, 17; SD6, 7; OV9, 10, 19, 22-24, 38;
Ell.:

L	F	R	N	S
7	6	6	6	0

Gri.: C/CR

Ulex europaeus (gorse)

Rough grassy places and heaths, usually on light impoverished soils (nitrogen fixing bacteria in root nodules)

Indicates: Moderately acid, more or less infertile sites e.g. sandy or peaty soil

References: Hill *et al.* (2004), Humphries & Shaughnessy (1987), Stace (2010)

Ph. 1: A Woodland and scrub, B1.1 Acid grassland – unimproved, B2.1 Neutral grassland – unimproved, B3.1 Calcareous grassland – unimproved, B3.2 Calcareous grassland – semi-improved, B5 Marsh/marshy grassland, D1.1 Dry dwarf shrub heath – acid, D1.2 Dry dwarf shrub heath – basic, D2 Wet dwarf shrub heath, C3.1 Tall ruderal;
NVC W10, 16, 22-25; M16, 25; H1-3, 5-8; MG5; CG2-5, 8; U1-3; OV27;
Ell.:

L	F	**R**	**N**	S
7	5	**5**	**3**	0

Gri.: SC

Ulmus procera (English elm)

An Ellenberg reaction value of 8 indicates a plant seldom found on acid soils

Indicates: Weakly basic to basic soils

Ph. 1: A Woodland and scrub;
NVC: W8, 21;
Ell.:

L	F	**R**	N	S
5	5	**8**	6	0

Gri.: C/SC

Umbilicus rupestris (navelwort)

A succulent of free-draining habitats with limited rooting opportunities and liable to intermittent water stress: acid rock, limestone, walls, thatch and on trees. Comparison of the distribution with annual precipitation over UK or February minimum isotherm suggests a possible link.

Speculative indicator: Annual average rainfall exceeds 800 mm or local hydrological conditions operating which compensate for a lower precipitation. Possible useful indicator of shifting climate envelopes in UK.

References: Amphlet & Rea (1909), Crawley (2005), Daniel *et al.* (1985), Edgington (2002), Gillam (1993), Grose (1957), Hedderson *et al.* (2003a), Preston *et al.* (2002), Sinker *et al.* (1991), Wade (1970), www.metoffice.gov.uk/

Ph. 1: B1.1 Unimproved acid grassland, I Rock exposure and waste, J2.5 Wall (also an epiphyte);
NVC: U1;
Ell.:

L	F	R	N	S
6	4	5	4	0

Gri.: S/CSR

Urtica dioica (common nettle)

Agriculturally elevated soil phosphorus levels can remain for millennia after inputs cease if not removed by harvesting. This is used archaeologically to locate sites of past human occupation. Ancient woods lack both phosphate and phosphate-loving plants (*U. dioica, Sambucus nigra* (elder)and *Galium aparine* (cleavers)).

Indicates: Nutrient enrichment, especially phosphates. Former human habitation.

References:Compton & Boone (2000), Eidt (1977), McLauchlan (2006), Rackham (1986), Sandor & Eash (1995)

Ph. 1: A Woodland and scrub, B1 Acid grassland, B2 Neutral grassland, B3 Calcareous grassland, B5 Marsh/marshy grassland, C3.1 Tall ruderal, F Swamp, marginal and inundation, H2 Saltmarsh, J1 Cultivated/disturbed land, J2.5 Wall;
NVC: W2, 3, 5-10, 12-14, 19, 21, 22, 24, 25; M27, 28; MG1, 2, 9-11; CG2, 4, 6; U1; S3-7, 12, 14, 15, 23-26, 28; SM28; SD6, 12, 18; OV3, 8-10, 12, 13, 15, 17, 19, 21, 23-28, 33, 41, 42;
Ell.:

L	F	R	**N**	S
6	6	7	**8**	0

Gri.: C

Viburnum lantana (wayfaring tree)

Presence in quantity with dogwood (*Cornus sanguinea*) and privet (*Ligustrum vulgare*) indicates a calcareous soil

Indicates: Calcareous soil

References: Lousley (1969)

Ph. 1: A Woodland and scrub, J2 Boundaries;
NVC: W8, 10, 12, 21;
Ell.:

L	F	R	N	S
7	5	7	5	0

Gri.: N/L

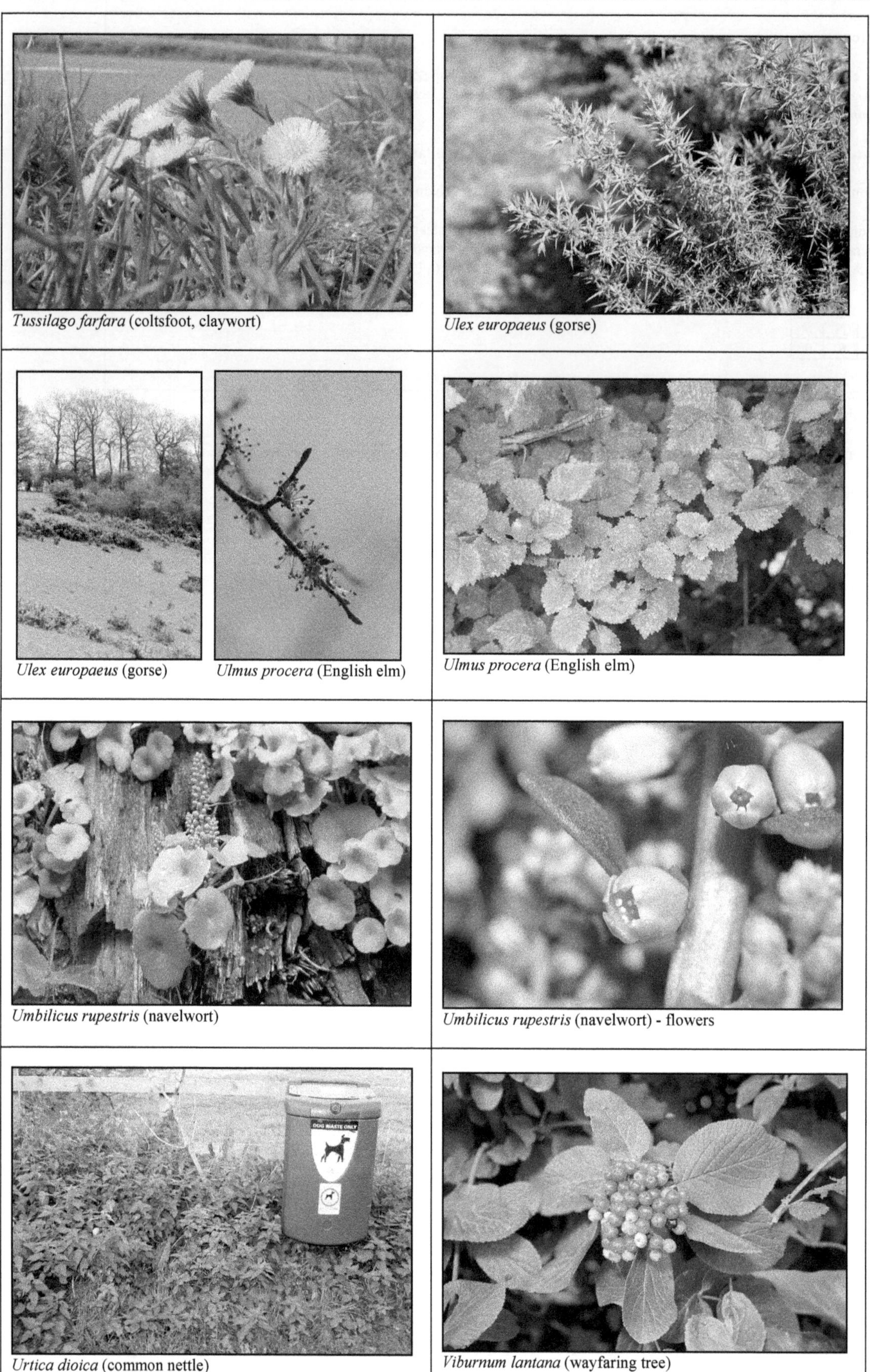

Tussilago farfara (coltsfoot, claywort)

Ulex europaeus (gorse)

Ulex europaeus (gorse)

Ulmus procera (English elm)

Ulmus procera (English elm)

Umbilicus rupestris (navelwort)

Umbilicus rupestris (navelwort) - flowers

Urtica dioica (common nettle)

Viburnum lantana (wayfaring tree)

Pteridophytes (ferns and allies)

Equisetum telmateia (giant horsetail)

Colonies are characteristic of sites where seeping, base-rich water emerges over heavy clay substrates.

Indicates: A spring or seepage line

References: Page (1988)

Ph. 1: E2 Flush and spring, F1 Swamp, F2.1 Marginal;
NVC: W5, 7; S18
Ell.:

L	F	R	N	S
6	8	7	6	0

Gri.: C/CSR

Ophioglossum vulgatum (adder's-tongue)

Characteristic of wet, oligotrophic grassland but also reported from moist woodland. Symbiotic, subterranean gametophyte. Damp soil profile probably important for completion of life cycle. Found in biodiverse swards in old meadows but also occurs in species-poor rush communities and road verges. Riddelsdell *et al.* (1948) consider it doesn't require 'wet' ground and include dry pastures among its habitats.

Indicates: Damp soil (not reliable for ancient, undisturbed grassland)

References: Castroviejo *et al.* (1986), Clausen (1938), Crowther *et al.* (2009), Muller (2000), Riddelsdell *et al.* (1948)

Ph. 1: B2.1 Neutral grassland – unimproved, B5 Marsh/marshy grassland, H6.4 Dune slack;
NVC: M22; MG5; SD14, 15;
Ell.:

L	**F**	R	**N**	S
8	**7**	7	**3**	0

Gri.: SR

Pteridium aquilinum (bracken)

Tends to be found on well-drained soils but not confined to them nor absent from waterlogged soil. Aeration is a limiting factor. Grows over a wide pH range although mostly found in moderately acidic soils.

Indicates:Suggestive of deep, well-drained loam or sandy loam

References: Marrs & Watt (2009), Poel (1948, 1951, 1961)

Ph. 1: A Woodland and scrub, B1 Acid grassland, B2 Neutral grassland, B3 Calcareous grassland – unimproved, B5 Marsh/marshy grassland, C1 Bracken, C3.1 Tall ruderal, D1.1 Dry dwarf shrub heath – acid, E1.6.1 Blanket bog, E1.7 Wet modified bog, H6.5 Dune grassland, H8.4 Coastal grassland;
NVC: W4, 7-11, 14-19, 21-23, 25; M15, 27; H1-4, 6, 8-10, 12, 21; MG1, 2, 5; CG10; U1-4, 20; SD9; MC12; OV2, 25, 27;
Ell.:

L	F	**R**	N	S
6	5	**3**	3	0

Gri.: C

Bryophytes (mosses and allies)

Campylopus introflexus (heath star-moss)

The presence of the alien bryophyte *Campylopus introflexus* on a bog surface or heathland is indicative of disturbance (e.g. trampling, burning, peat cutting), allowing this invasive southern hemisphere species to gain a purchase on the substrate and spread. It can arrive by spores or fragments. Considered the moss equivalent of *Rhododendron ponticum* in its invasive nature.

Indicates: Disturbance of acid soil/peat surface

References: Porley & Hodgetts (2005), Smith (2001), Söderström (1992)

Ph. 1: B1.1 Acid grassland – unimproved; H6.6 Dune heath; Dry heath/acid grassland (also on rotten wood and disturbed mires)
NVC: H4, H11; U1;
Ell.:

L	F	**R**	N	S	HM
7	5	**2**	2	0	2

Gri.: N/L

Ctenidium molluscum (chalk comb-moss)

A nitrophobe, avoiding high nitrogen environments; varieties found in many calcareous habitats can indicate base-rich conditions but var. *sylvaticum* is fairly frequent on acidic soil.

Indicates: low soil nitrogen status, usually base-rich

References: Atherton *et al.* (2010), Pitcairn *et al.* (2006)

Ph. 1: A Woodland and scrub, B3.1 Calcareous grassland – unimproved, B5 Marsh/marshy grassland, C2 Upland species-rich ledges, D4 Montane heath/dwarf herb, E2.2 Basic flush, E2.3 Bryophyte dominated spring, E3.2 Basin mire, I Rock exposure and waste including I1.3 Limestone pavement;
NVC: W8, 9, 13, 20; M8-11, 14, 26, 37, 38; CG2-14; U14, 15, 17; OV38, 40;
Ell.:

L	F	**R**	**N**	S	HM
7	6	7	**2**	0	1

Gri.: N/L

Fontinalis antipyretica (Greater water-moss)

Sensitive to pollution and grows well only in clean water. Most frequent in neutral or basic, mesotrophic or eutrophic water.

Indicates: 'Unpolluted' water

References: Hill *et al.* (1994), Porley & Hodgetts (2005)

Ph. 1: G1.1 Standing water – eutrophic, G1.2 Standing water – mesotrophic, G1.3 Standing water – oligotrophic, G1.5 Standing water – marl, G2.1 Running water – eutrophic, G2.2 Running water – mesotrophic, G2.3 Running water – oligotrophic, G2.5 Running water – marl, F1 Swamp;
NVC: A8, 11, 13, 14, 18, 19; S19;
Ell.:

L	**F**	R	**N**	S	**HM**
6	**12**	6	**5**	0	**0**

Gri.: N/L

Funaria hygrometrica (common cord-moss) - 'bonfire moss'.

Grows on recently burned areas and can form dense stands in the first year after a fire when soil nutrient levels are elevated. These stands decline as other colonists arrive and out-compete them. Occurs on disturbed sites other than burned areas (e.g. felled woodland). It appears after fire in many countries.

Indicates: Site of a fire (e.g. bonfire), recently felled woodland or other disturbed sites. Look for charcoal in the soil beneath.

References: Hoffman (1966), Lepp (2008)

Ph. 1: J1.1 Arable, J1.3 Ephemeral/short perennial;
NVC: OV19, 22;
Ell.:

L	F	R	**N**	S	HM
7	5	6	7	0	1

Gri.: N/L

Pteridophytes (ferns and allies)

Equisetum telmateia (giant horsetail)

Ophioglossum vulgatum (adder's-tongue)

Pteridium aquilinum (bracken fronds)

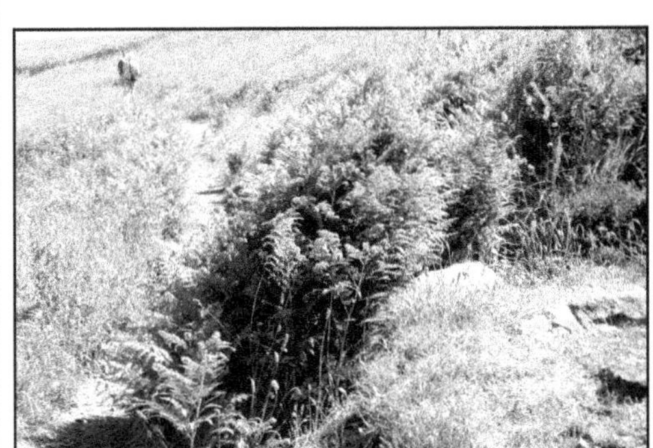
Pteridium aquilinum (bracken colony)

Bryophytes (mosses and allies)

Campylopus introflexus (heath star-moss)

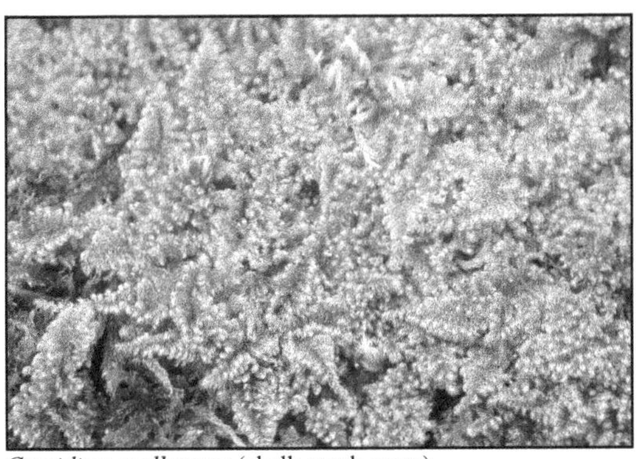
Ctenidium molluscum (chalk comb-moss)

Fontinalis antipyretica (Greater water-moss)

Funaria hygrometrica (common cord-moss) – 'bonfire moss'

Polytrichum piliferum (bristly hair-cap)

Pioneer species of dry, acidic early successional habitats such as bare acid sand, burnt moorland, coal waste and disturbed ground.

Indicates: Disturbed soil surface

References: Atherton *et al.* (2010), Miller *et al.* (2007)

Ph. 1: B1.1 Acid grassland – unimproved, D1 Dry dwarf shrub heath, D4 Montane heath/dwarf herb, H6.6 Dune heath, H6.8 Open dune, I Rock exposure and waste;
NVC: H2, 8, 10, 11, 14, 16, 19, 20; U1, 2, 7, 9, 10, 12, 14, 18; SD11;
Ell.:

L	F	R	N	S	HM
9	3	3	1	0	2

Gri.: N/L

Ptilidium ciliare (ciliated fringewort)

A species of usually well-drained, infertile, acidic substrates including coal waste.

Indicates: Low nutrient, acid environment including coal waste

References: Atherton *et al.* (2010), Miller *et al.* (2007)

Ph. 1: A Woodland and scrub, B1.1 Acid grassland – unimproved, B3.1 Calcareous grassland – unimproved, C1 Bracken, C3.2 Tall herb and fern – non-ruderal, D1.1 Dry dwarf shrub heath – acid, D4 Montane heath/dwarf herb, D5 Dry heath/ acid grassland, E1.6.1 Blanket bog, E1.6.2 Raised bog, E3.2 Basin mire, H6.6 Dune heath, H6.8 Open dune, I Rock exposure and waste;
NVC: W17, 20; M7, 8, 17, 19, 20; H1, 9, 11-14, 16, 18-22; CG7, 10-12, 14; U1, 5-10, 13, 14, 18-21; SD11;
Ell.:

L	F	R	N	S	HM
6	5	2	2	0	0

Gri.: N/L

Rhytidiadelphus squarrosus (springy turf-moss)

A nitrophobe. Refer to *Hygrocybe* species (waxcap fungi).

Indicates: Rarely abundant in new lawns. If present in quantity, check for emergence of waxcap (*Hygrocybe* species) fungi at appropriate times of year.

References: Griffith *et al.* (2002), Pitcairn *et al.* (2006)

Ph. 1: A Woodland and scrub, B5 Marsh/marshy grassland, E1 Bog, E3.2 Basin mire, E2.1 Acid/neutral flush, E2.2 Basic flush, E2.3 Bryophyte dominated spring;
NVC: W1, 4, 9, 11, 17, 19, 20, 23; M5, 6, 8, 12, 15, 17, 19, 23, 25-28, 32, 35; H4, 8, 10, 12, 16, 18, 19; MG1-3, 5, 6, 9; CG2, 4, 6, 8-14; U1-7, 13-21; S1; SM16; SD7-9, 12, 17; MC9; OV37;
Ell.:

L	F	R	N	S	HM
7	5	5	4	0	2

Gri.: N/L

Sphagnum species (bog mosses)

The species show a range of tolerance to pollution and habitat disturbance. The composition of the assemblage can therefore indicate habitat quality.

Indicates: Bog mosses typify an acidic habitat with impeded drainage

References: Hakan *et al.* (2006)

Ph. 1: A1 Woodland and scrub, B1 Acid grassland, C3.1 Calcareous grassland – unimproved, C3.2 Tall herb and fern – non-ruderal, C2 Upland species-rich ledges, D4 Montane heath/dwarf herb, D5 Heathland, H6.6 Dune heath, E Mire, F Swamp, marginal and inundation;
NVC: W2-5, 7, 11, 17, 18, 20; M1-10, 12, 14-21, 24, 25, 29-35; H10, 14, 15, 18, 20-22, U5-8, 13, 16-18, 21, CG14; S2, S4, S27;
Ell.:

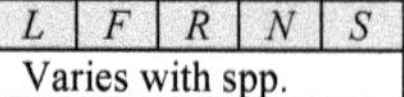

L	*F*	*R*	*N*	*S*
Varies with spp.				

Gri.: N/L

Algae
Batrachospermum sp. (a freshwater red alga) A conspicuous species of cold running streams. It is a common, macroscopic alga widely reported from springs. **Indicates: Extremely pure, katharobic conditions** (oxygenated and low in organic matter such as occur in a limestone spring) References: Round (1973) Ph. 1: E2 Flush and spring; G2.3 Running water – oligotrophic NVC: N/L; Ell.: N/L; Gri.: N/L
Charophytes (stoneworts – green, freshwater algae) Genera: *Chara, Lamprothamnium, Nitellopsis, Nitella, Tolypella.* Lakes dominated by these are typically hard water, calcium-rich and low in phosphate. Majority tolerate pH 6 to pH 9 (some *Nitella* species from pH 5). *Chara* mostly from mesotrophic/eutrophic waters, *Nitella* mostly from oligotrophic waters. Phosphate levels > 20 μgl^{-1} appear to inhibit charophytes. **Indicates: Hard water, calcium-rich, low phosphate (early successional)** References: Forsberg (1965), Moore (1986) Ph. 1: G Standing water; NVC: A7, 8, 11, 13, 14, 23, 24; Ell.: N/L; Gri.: N/L
Ulva (Enteromorpha) intestinalis (sea-guts) A green, marine alga found at all levels of the shore, but able to withstand low salinities and thriving in areas where freshwater run-offs occur. Usually grows as a mass of tubular thalli but sometimes proliferates in 'bottle-brush' form. Proliferous forms may be more common in areas with an unstable salinity regime (e.g. estuaries). Also an indicator of eutrophication. **Indicates: Brackish conditions and eutrophication** References: Fish & Fish (2011), Reed & Russell (1978) Ph. 1: G Open water (brackish); NVC: N/L; Ell.: N/L; Gri.: N/L

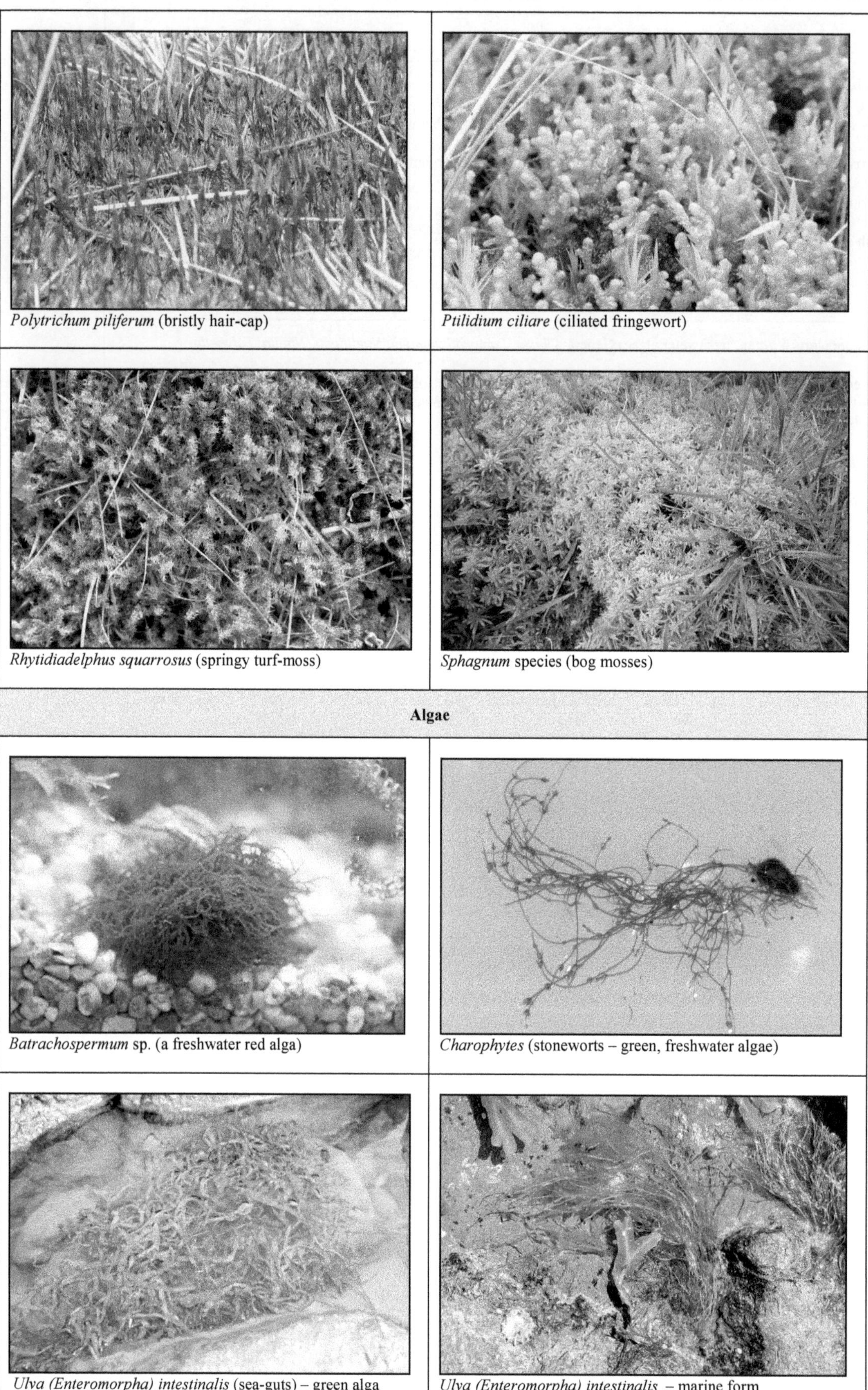

Polytrichum piliferum (bristly hair-cap)

Ptilidium ciliare (ciliated fringewort)

Rhytidiadelphus squarrosus (springy turf-moss)

Sphagnum species (bog mosses)

Algae

Batrachospermum sp. (a freshwater red alga)

Charophytes (stoneworts – green, freshwater algae)

Ulva (Enteromorpha) intestinalis (sea-guts) – green alga

Ulva (Enteromorpha) intestinalis – marine form

Fungi
Calvatia gigantea (giant puffball) *Calvatia gigantea* is a nitrophilous fungus, found in nutrient-rich grasslands, parks, cultivated fields, compost heaps, hedgerows and woodland. **Indicates: Nutrient-rich soil** References: Pegler *et al.* (1995) Ph. 1: A1 Woodland, B4 Improved grassland; J1 Cultivated land, J2 Boundary; NVC: N/L; Ell.: N/L; Gri.: N/L
Chlorociboria aeruginascens and *C. aeruginosa* (green elfcup) Fragments of branches and rotten wood (in deciduous woodlands) showing a distinctive turquoise blue colour are probably this. Formerly used in the manufacture of Tunbridge ware. Fence posts showing green colour may be impregnated with preservative. **Indicates: broad-leaved wood ('hard wood')** Reference: Ramsbottom (1953) Ph. 1: A Woodland; NVC: N/L.; Ell.: N/L; Gri.: N/L
Hygrocybe species (waxcap fungi) Diverse waxcap populations associated with regularly grazed or mown grasslands with no recent application of fertilizer. Also associated with bryophytes e.g. *Rhytidiadelphus squarrosus* and *Pseudoscleropodium purum*. *Hygrocybe* spp. can appear in highly anthropogenic habitats such as turf roofs. **Indicates: Regular grazing/mowing, no recent fertiliser application** References: Arnolds (1981, 1982), Griffith *et al.* (2002) Ph. 1: B Grassland (unimproved): NVC: N/L; Ell.: N/L; Gri.: N/L
Rhytisma acerinum (tar spot fungus on sycamore leaves) The distribution of tar spot is limited by a concentration of aerial sulphur dioxide between 80-90 μgm^{-3}. High numbers of spots may indicate relatively clean air but presence of infected leaf litter increases likelihood of reinfection of new leaves in spring, therefore interpretation requires care. Fewer spots may indicate polluted air or absence of fungal spores due to clearing of dead leaves in autumn by man or wind. **Indicates: When present in quantity, sulphur dioxide levels may be relatively low** References: Ellis & Ellis (1985), Leith & Fowler (1987), Vick & Bevan (1976) Ph. 1: A1 Woodland, A3 Parkland/scattered trees; NVC: N/L; Ell.: N/L; Gri.: N/L

Lichenised fungi (lichens)
Caloplaca spp. (orange crustose lichen) Inland, basic substrates are colonized by orange crustose *Caloplaca* spp. Some yellow/orange *Caloplaca* spp. occur on acid substrates inland but do not usually produce an unbroken crust. The orange thallus of *Xanthoria* is foliose (a fingernail can be inserted beneath the edge). NB: on coastal sites some orange *Caloplaca* spp. occur on siliceous rock. **Indicates: Basic rock** References: Dobson (2005) Ph. 1: I1.1.2 Inland cliff, I2.2.2 Scree – basic, I1.3 Limestone pavement, J2.5 Wall; NVC: N/L; Ell.: N/L; Gri.: N/L
Cladonia species (various) *Cladonia* species are brittle, fruticose lichens that sometimes form large mats on the ground. This lichen heath is an uncommon habitat but it can form on anthropogenic substrates such as coal waste, limestone pipe-bedding and road verges. If present it is worth searching for unusual lichens and other small species. **Indicates: Presence in quantity at least indicates a lack of trampling** Ph. 1: A Woodland and scrub, B1.1 Acid grassland – unimproved, B3.1 Calcareous grassland – unimproved, C1 Bracken, C3.1 Tall herb and fern – non-ruderal, D Heathland, D3 Lichen/bryophyte heath, E Mire, H4 Rocks/boulders above high-tide mark, H6.5 Dune grassland, H6.6 Dune heath, H6.7 Dune scrub, H8.3 Crevice/ledge vegetation, H8.4 Coastal grassland, I Rock exposure and waste, J2.5 Wall; NVC: W15-20; M2, 7, 12, 15-21; H1-4, 6-22; CG1, 2, 7, 9, 11-14; U1, 2, 3; 5-14, 16, 18-21; MC5, 9; SD7, 8, 11, 12, 18; OV 37-39; Ell.: N/A; Gri.: N/A
Cladonia cervicornis subspecies *verticillata* (resembles small, fairytale castles) **Indicates: Associated with heavy metal contamination and/or coal waste** References: Gilbert (2000), Seaward (1996) Ph. 1: D2 Wet dwarf shrub heath, D3 Lichen/bryophyte heath, D6 Wet heath/acid grassland, H6.6 Dune heath, I2 Artificial exposures and waste; NVC: M16; Ell.: N/A; Gri.: N/A
Peltigera species (dog-lichens) High growth rates and photosynthetic capacity and nitrogen fixation by symbiotic blue-green algae enable dog-lichens to exploit nutrient-poor substrates. **Indicator: Nutrient-poor substrate, primary colonizer** References:Hahn *et al.* (1993), Palmqvist & Sundberg (2000), Schell & Alexander (1973), Sundberg *et al.* (2001),Webster & Brown (1997) Ph. 1: A Woodland and scrub, B1.1 Acid grassland – unimproved, B2.1 Neutral grassland – unimproved, B3.1 Calcareous grassland – unimproved, D4 Montane heath/dwarf herb, C2 Upland species-rich ledges, C3.2 Tall herb and fern – non-ruderal, H6.5 Dune grassland, H6.6 Dune heath, H8.4 Coastal grassland, I Rock exposure and waste; NVC: W20; H11; MG2; CG7, 10-12, 14; U1, 10, 13, 17, 18; SD7, 8, 12, 19; MC9; Ell.: N/L; Gri.: N/L

Fungi	
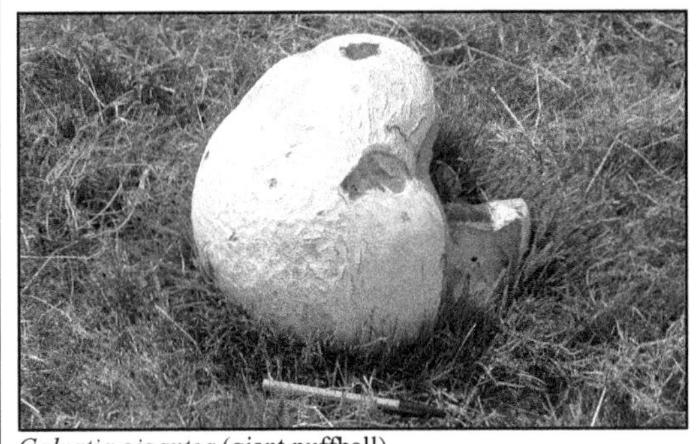 *Calvatia gigantea* (giant puffball)	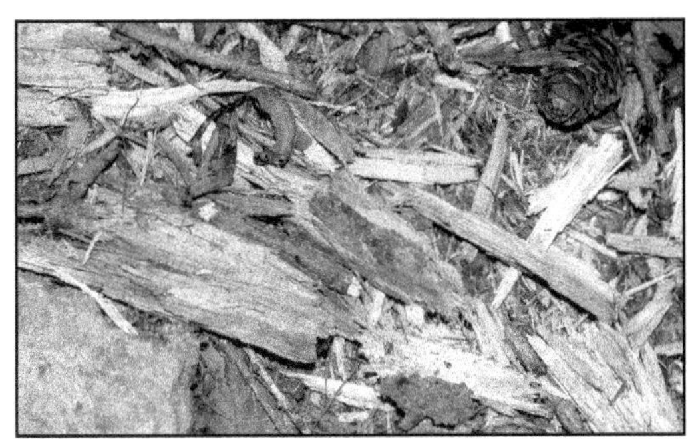 *Chlorociboria aeruginascens* and *C. aeruginosa* (green elfcup)
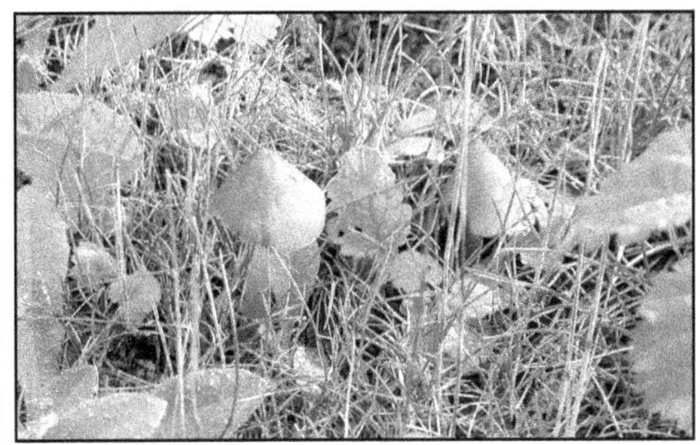 *Hygrocybe* species (waxcap fungi)	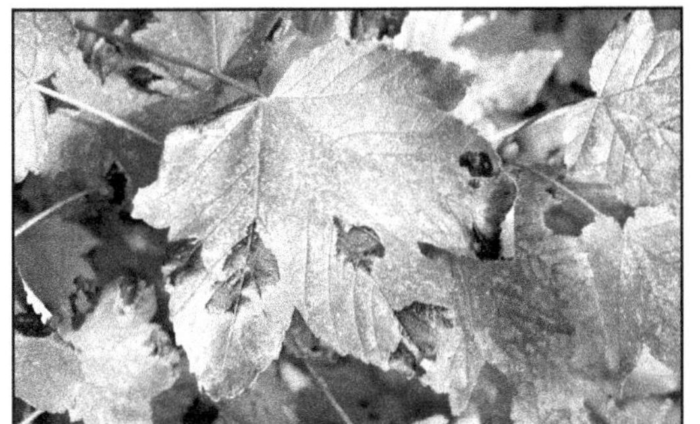 *Rhytisma acerinum* (tar spot fungus on sycamore leaves)
Lichenised fungi (lichens)	
 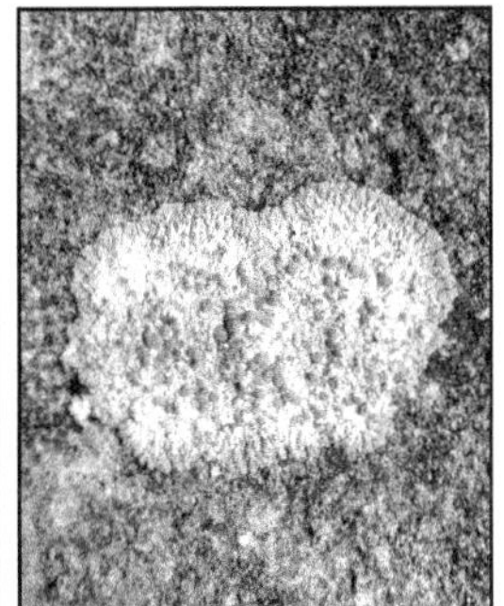 *Caloplaca* spp. (orange crustose lichen)	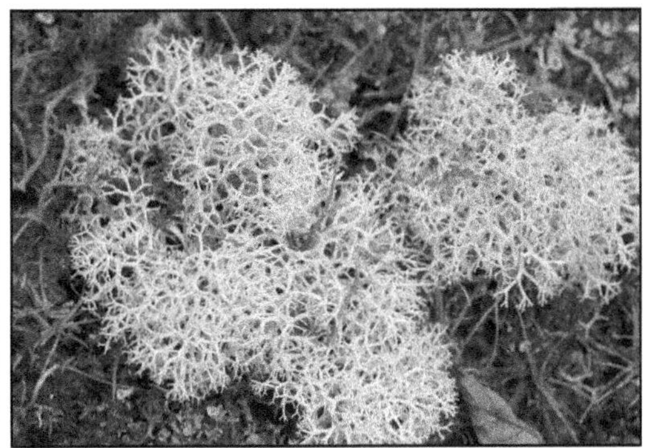 *Cladonia* species (various)
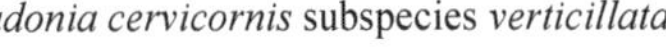 *Cladonia cervicornis* subspecies *verticillata*	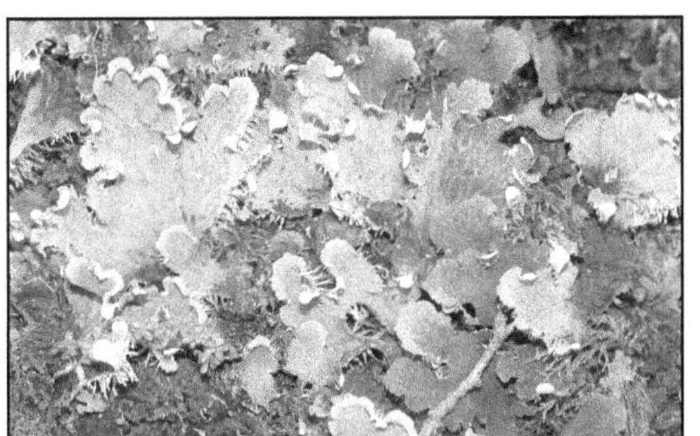 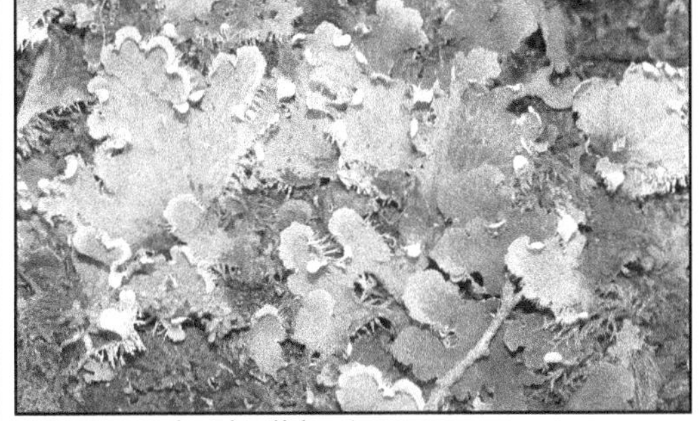 *Peltigera* species (dog-lichens)

Rhizocarpon geographicum (map lichen)

A lichen of acid rock faces and some urban walls and slate roofs. The distinctive lime-green thallus with black lines is highly recognizable.

Indicates: Green, crustose lichen on rock indicates acid rock

References: Gilbert (2000)

Ph. 1: I Rock exposure and waste;
NVC: N/L;
Ell.: N/L;
Gri.: N/L

Stereocaulon species (a lichen group resembling sugar-frosting)

A genus of upland siliceous rock, including recent volcanic rock, metal-rich spoil heaps and other substrates. It has been speculated that the appearance of *Stereocaulon* species in towns coincided with the rise in car use since World War II and the use of leaded petrol.

Indicates: Consider a heavy metal influence

References: Gilbert (2000), Smith *et al.* (2009)

Ph. 1: D4 Montane heath/dwarf herb, I Rock exposure and waste;
NVC: U9;
Ell.: N/L;
Gri.: N/L

Usnea species (lichens on trees resembling grey beards)

A 'clean-air' lichen. *U. subfloridana* is the first *Usnea* species to return to an area if air quality improves. It disappeared from many areas before 1970 but is now returning as sulphur dioxide levels fall in response to Clean Air Acts.

Indicates: Presence indicates 'clean air' (but difficult to know whether it is improving or deteriorating without monitoring data)

References: Dobson (2005)

Ph. 1: A Woodland and scrub (including A3 Parkland/scattered trees as an epiphyte), H6.6 Dune heath (on ground);
NVC: H11;
Ell.: N/L;
Gri.: N/L

Xanthoria parietina (golden shield lichen)

Thrives on nutrient-rich substrates and is becoming more common due to nitrification of the environment.

Indicates: Site of nutrient enrichment, e.g. bird guano, fertilizer drift

References: Smith *et al.* (2009)

Ph. 1: A Woodland and scrub (including A3 Parkland/scattered trees, as an epiphyte), I Rock exposure and waste, J2 Boundary, J3 Building;
NVC: N/L;
Ell.: N/L;
Gri.: N/L

Bacteria

Nostoc commune (cyanobacteria, blue-green algae, star jelly, witches' butter and mares' eggs)

Brown jelly on urban paths/lawns. Widely distributed on alkaline soils/moist rock, it can remain desiccated for years and recover metabolic activity within hours after rehydration. Withstands repeated freezing/thawing. An important component of extreme terrestrial arctic/Antarctic habitats and able to fix atmospheric nitrogen.

Indicates: Recent moisture (becomes conspicuous after rain, raising consternation amongst gardeners fearing a threat to lawns)

References: Bold & Wynne (1978), Dodds *et al.* (1995)

Ph. 1: I2 Artificial exposures and waste tips, J1 Cultivated/disturbed land (sometimes in amenity grassland);
NVC: N/L;
Ell.: N/L;
Gri.: N/L

 *Rhizocarpon geographicum* (map lichen)	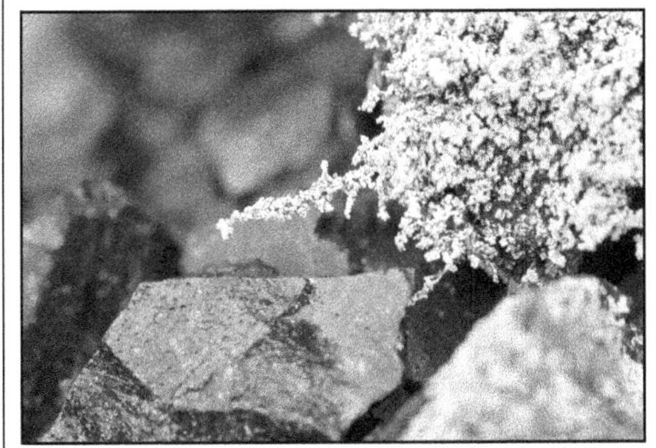 *Stereocaulon* species (lichens resembling sugar-frosting)
 *Usnea s*pecies (lichens resembling grey beards, on trees)	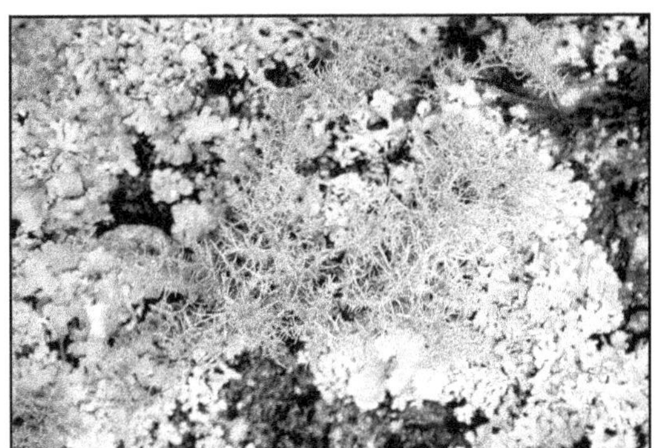 Fruticose *Usnea s*pecies with foliose lichens
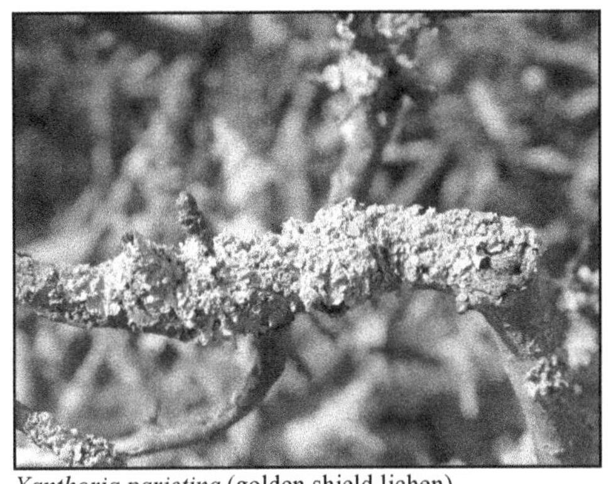 *Xanthoria parietina* (golden shield lichen)	 Proliferation of *Xanthoria parietina* on Sewage filter bed
Bacteria	
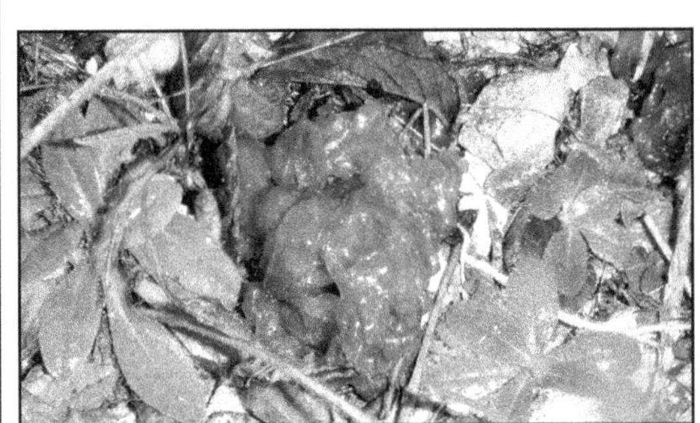 *Nostoc commune* (cyanobacteria, blue-green algae)	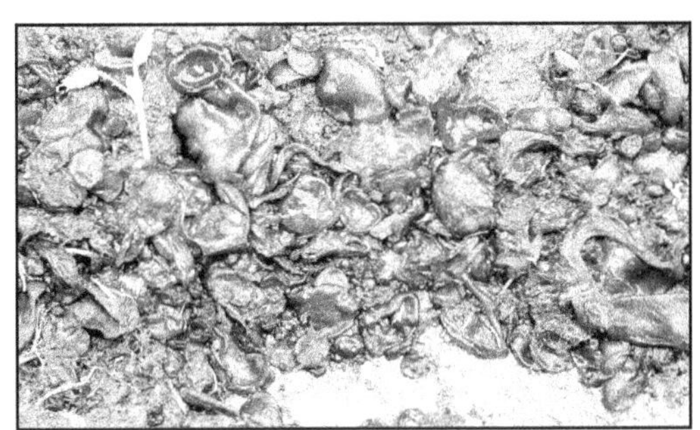 *Nostoc* colony beside a path

TABLE OF INDICATORS 2: HABITATS AND FEATURES

DEDUCING SOIL TYPE

a) Identifying acidic, basic and neutral (mesotrophic) soils from the plant species present

Indicators: Know the **calcicole** ('calcium loving') and **calcifuge** ('calcium avoiding') plants.

Calcium carbonate (chalk) is a soluble base and neutralises soil acidity. Soils low in calcium carbonate or other bases are acidic. Calcicole plants are associated with calcium carbonate-rich soil; often those overlying chalk or limestone. Calcifuges avoid such soils. A pH above 6.5 limits the vigorous growth of many calcifuges, **lime-induced chlorosis** (yellowing) and other adverse effects occurring above this. Other soluble bases such as calcium magnesium carbonate may also reduce soil acidity (Refer to Clapham *et al.*,1962 and Grime, 1963).

Acid soils *Calluna vulgaris* (heather), *Cuscuta epithymum* (dodder), *Deschampsia flexuosa* (wavy hair-grass), *Digitalis purpurea*(foxglove),*Drosera rotundifolia* (round-leaved sundew), *Erica* species(heaths), *Galium saxatile* (heath bedstraw),*Jasione montana*(sheep's-bit scabious),*Narthecium ossifragum* (bog asphodel), *Pinus sylvestris* (Scot's pine), *Rumex acetosella* (sheep's sorrel), *Ulex* species(gorse), *Vaccinium myrtillus* (bilberry) **Indicate: acid soils** References:Hill *et al.* (2004), Lousley (1969), Steele (1955)
Calcareous (basic) soils *Asperula cynanchica* (squinancywort), *Blackstonia perfoliata* (yellow-wort), *Brachypodium pinnatum* (tor grass), *Carduus nutans* (nodding thistle), *Centaurea scabiosa* (greater knapweed), *Ceterach officinarum* (rusty-backfern), *Chelidonium majus* (greater celandine), *Cirsium acaule* (stemlessthistle), *Cirsium eriophorum* (woolly thistle), *Clematis vitalba* (travellers' joy), *Euonymus europaeus* (spindle), *Helleborus foetidus* (stinking hellebore), *Helleborus viridis* (green hellebore), *Hippocrepis comosa* (horseshoe vetch), *Inula conyzae* (ploughman's spikenard), *Linaria vulgaris* (common toadflax), *Ophrys apifera* (bee orchid), *Orobanche minor* (common broomrape), *Parietaria judaica* (pellitory-of-the-wall), *Pseudofumaria lutea* (yellow corydalis), *Resdea luteola* (weld), *Sanguisorba minor* (salad burnet), *Scabiosa columbaria* (small scabious), *Silene vulgaris* (bladder campion), *Thymus pulegioides* (large thyme), *Ulmus procera* (English elm), *Viola hirta* (hairy violet) **Indicate: calcareous soils** References:Hill *et al.* (2004), Lousley (1969), Steele (1955)
Mesotrophic (neutral) soils There are few diagnostic, indicator species for neutral grassland. It tends to lack strong calcicoles or calcifuges. The following list of characteristic species is compiled from four published sources and the author's experience. *Ajuga reptans* (bugle), *Anthoxanthum odoratum* (sweet vernal grass), *Cardamine pratensis* (cuckoo flower), *Cynosurus cristatus* (crested dog's-tail), *Galium verum* (lady's bedstraw), *Lathyrus pratensis* (meadow vetchling), *Leontodon autumnalis* (autumn hawkbit), *Leontodon hispidus* (rough hawkbit), *Leucanthemum vulgare* (oxeye daisy), *Lotus corniculatus* (bird's-foot trefoil), *Primula veris* (cowslip), *Ranunculus acris* (meadow buttercup), *Rhinanthus minor* (yellow-rattle), *Trifolium pratense* (red clover) **Indicate: Neutral grassland** References: Bedfordshire Wildlife Trust (2009), Riddelsdell *et al.* (1948), Rural Development Service (2005), Tansley (1949)

a) Identifying soil nutrient status from the plant species present

Global energy use and food production have increased nitrogen (i.e. compounds of nitrogen[6]) inputs to ecosystems worldwide, impacting plant community diversity, composition, and function (Clark *et al.*, 2007). Studies frequently find species richness declines with nitrogen enrichment and increasing primary production. Nitrogen enrichment can lead to gains and losses of species, changes in dominance and rarity, and shifts in relative species abundance. Nitrogen has traditionally been considered the primary limiting nutrient for plant growth in terrestrial ecosystems, but recent work suggests that phosphorus, water and other resources are also involved (Cleland & Harpole, 2010). Knowledge of the nutrient status and relative moistness of habitats is

[6] In this section the elemental names 'nitrogen' and phosphorus' are used to mean the compounds of these elements as they occur naturally in soil and water (e.g. ammonia, ammonium compounds, nitrites, nitrates, phosphates).

useful when planning ecological management or translocation protocols. Indicator plants can provide some subjective insight into the nutrient status and moistness of sites.

Indicators: Know the **nitrophile** ('nitrogen loving') and **nitrophobe** ('nitrogen avoiding') plants, and those whose persistence is related to phosphate levels, and gain an insight into the relative moisture dependencies of different terrestrial plant species. Examples of indicator species useful in assessing these environmental parameters are provided below.

Nutrient-rich soils

Species associated with nutrient-rich soils.

Indicators of nutrient-rich situations: *Atriplex prostrata* (spear-leaved orache), *Epilobium hirsutum* (great willowherb), *Stellaria media* (common chickweed), *Typha latifolia* (reedmace), *Beta vulgaris* (beet), *Galium aparine* (cleavers), *Lamium album* (white dead-nettle), *Urtica dioica* (common nettle)

Indicators of extremely nutrient-rich sites (e.g. cattle resting places): *Arctium lappa* (greater burdock), *Artemisia absinthium* (wormwood), *Hyoscyamus niger*(henbane), *Rumex obtusifolius* (broad-leaved dock)

Indicate: Nutrient-rich soils

References: Hill *et al.* (1999)

Infertile soils

Species associated with more or less infertile sites

Centaurea scabiosa (greater knapweed), *Galium saxatile* (heath bedstraw), *Pimpinella saxifraga* (burnet-saxifrage), *Teucrium scorodonia* (wood sage), *Aira praecox* (early hair-grass), *Carex panicea* (carnation sedge), *Linum catharticum* (fairy flax), *Scabiosa columbaria* (small scabious)

Species associated with extremely infertile sites

Agrostis curtisii (bristle bent), *Clinopodium acinos* (basil thyme), *Drosera rotundifolia* (round-leaved sundew) and *Rubus chamaemorus* (cloudberry)

Indicate: Species associated with infertile soils

References: Hill *et al.* (1999)

Nitrophiles: Plants thriving in high nitrogen environments

Agrostis stolonifera (creeping bent), *Anthriscus sylvestris* (cow parsley), *Arrhenatherum elatius* (false oat-grass), *Brachypodium pinnatum* (tor grass), *Chamerion angustifolium* (rosebay willowherb), *Dactylis glomerata* (cock's-foot), *Dryopteris dilatata* (broad buckler fern), *Festuca rubra* (creeping fescue), *Galium aparine* (cleavers), *Geum urbanum* (wood avens), *Glechoma hederacea* (ground ivy), *Hedera helix* (ivy), *Kindbergia praelonga* (common feather-moss), *Plantago major* (great plantain), *Rubus fruticosus* (bramble), *Rumex obtusifolius* (broad-leaved dock), *Sambucus nigra* (elder), *Stellaria media* (common chickweed), *Urtica dioica* (common nettle)

Indicate: High soil nitrogen

References: Bobbink & Lamers (2002), Pitcairn *et al.* (2006)

Nitrophobes: Plants 'avoiding' high nitrogen environments

Calluna vulgaris (heather), *Campanula rotundifolia* (harebell), *Carex flacca* (glaucous sedge), *Cladonia* spp. (pixie-cup and allied lichens), *Ctenidium molluscum* (chalk comb-moss), *Erica* spp. (heaths), *Frullania tamarisci* (tamarisk scalewort), *Helianthemum nummularium* (common rock-rose), *Hieracium pilosella* (mouse-ear hawkweed), *Potentilla erecta* (tormentil), *Ptilidium ciliare* (ciliated fringewort), *Rhytidiadelphus squarrosus* (springy turf-moss), *Sphagnum* spp. (bog-mosses), *Succisa pratensis* (devil's-bit scabious), *Vaccinium myrtilus* (bilberry)

Indicate: Low soil nitrogen

References: Bobbink & Lamers (2002), Pitcairn *et al.* (2006)

Acid soils

Acid wall community, Lizard, 19 June

Calcareous (basic) soils

Limestone grassland community, WhitePeaks, 13 May

Mesotrophic (neutral) soils

Road verge community, Ettington, 11 June

Mesotrophic (neutral) soils

Flood meadow community, Cricklade, 19 April

Nutrient-rich soils

Galium aparine (cleavers)community, Gloucestershire, 31 May

Infertile soils

Campanula rotundifolia (harebell) community, Worcestershire, 1 August

Nitrophiles: Plants thriving in high nitrogen environments

Tall ruderal, nitrophile community, Gloucestershire, 5 June

Nitrophobes: Plants 'avoiding' high nitrogen environments

Heathland, nitrophobe community, Exmoor, 10 August

c) Identifying variation in soil moistness from the plant species present

Some plants are restricted to wetland habitats but 'wet' is a relative term and the soil moisture dependency of different plant species is also relative. Plant species which may give some clues with regard to relative moistness of soils in terrestrial (non-wetland) habitats are indicated below.

Moist site indicators

Plants typical of moist (not wet) soils.

Anthriscus sylvestris (cow parsley), *Euphorbia amygdaloides* (wood spurge), *Hyacinthoides non-scripta* (bluebell), *Solanum nigrum* (black nightshade)

Indicate: Moist soil

References: Hill *et al.* (1999)

Dry site indicators

Plants typical of dry soils.

Asplenium trichomanes (maidenhair spleenwort), *Centaurea scabiosa* (greater knapweed), *Spergularia rubra* (sand spurrey), *Clinopodium acinos* (basil thyme), *Saxifraga tridactylites* (rue-leaved saxifrage)and *Sedum acre* (biting stonecrop).

Indicators of extreme dryness (restricted to soils that often dry out for some time) include *Helianthemum apenninum* (white rock-rose) and *Koeleria vallesiana* (Somerset hair-grass)

Indicate: Dry soil

References: Hill *et al.* (1999)

DEDUCING CONDITION OF A WETLAND

Just as for terrestrial ecosystems, wetland habitats are vulnerable to contamination with excess nutrients. This may lead to eutrophication and the loss of diversity through the selective blooming of nitrogen or phosphate demanding species but many other forms of pollution also arise and affect wetland communities in different ways. Some simple rules of thumb of use when making superficial assessments of wetland quality are provided in the following section. These are not intended as an alternative to rigorous chemical or ecological investigation of water quality but merely as coarse tools providing hints to the nature of underlying conditions.

The adjectives 'polluted' and 'unpolluted' are highly unspecific and, used in isolation, offer little objective information when describing a wetland. Nevertheless, these terms often run through the surveyor's mind when working in the field and comparing aquatic habitats. A wetland is often referred to as 'polluted' when an observer considers that its ecological quality in some way conflicts with expectations. Some of the ecological indicators used in the subjective assignment of this suboptimal state are provided below.
It should be noted that, not all apparent wetland 'pollution' phenomena are man-made or harmful and a few common examples are provided. Orange precipitates, surface films and foams in, or on, watercourses are often observed by field surveyors. These are sometimes the result of man-made pollution but often they are the consequence of other processes and may be ecologically benign.

A nitrophilous marsh community

Presence of the nitrophiles *Epilobium hirsutum* (great willowherb), *Urtica dioica* (common nettle) and *Solanum dulcamara* (woody nightshade)suggest a eutrophicated situation.

Indicate: Eutrophicated wetland

References: Gilbert (1991)

'Polluted' and 'unpolluted' water

The most polluted sites have no macrophytes (visible aquatic plants). As water quality improves, tolerant species like *Potamogeton pectinatus* (fennel pondweed)and blanket weed (filamentous algae) may appear. Where pollution is moderate, different pollutants may have different effects. Aquatic *Ranunculus* species (water buttercups) appear in 'unpolluted' situations.

Indicate: Greater range of visible aquatic plants indicates less pollution

References: Haslam (1990)

Eutrophicated (nutrient-rich) water

Some water plants which thrive in eutrophic or nutrient-rich water.

Lemna gibba (fat duckweed), *Azolla filiculoides* (water fern), *Ceratophyllum demersum* (rigid hornwort), *Ceratophyllum submersum* (soft hornwort), *Ranunculus aquatilis* (common water-crowfoot), *Rumex hydrolapathum* (water dock), *Rorippa amphibia* (great yellow-cress), *Butomus umbellatus* (flowering-rush), *Sagittaria sagittifolia* (arrowhead), *Glyceria maxima* (reed sweet-grass), *Glyceria fluitans* (floating sweet-grass), *Typha latifolia* (reedmace)

Indicate: Eutrophic water

References: Preston & Croft (1997)

Oligotrophic (nutrient-poor) water

Some water plants which are characteristic of oligotrophic water.

Isoetes lacustris (quillwort), *Lobelia dortmanna* (water lobelia). The fact that these species are generally, less-frequently encountered by ecologists than those listed under 'Eutrophicated water' above indicates something about the state of the UK's wetlands.

Indicate: Oligotrophic water

References: Preston & Croft (1997)

Orange precipitates in streams

Orange precipitates of ochre on rocks, twigs and roots in streams or springs (chalybeate or iron-rich springs). These are formed either by the action of a diverse group of microorganisms known as 'iron bacteria', when they are knownas 'slimy flocculates', or by abiotic processes.

Ochre is hydrated ferric oxide. In anaerobic conditions, iron in water is reduced to give soluble ferrous salts. When a solution of such salts emerges into a well-aerated stream, they are oxidized to ferric salts by abiotic or biotic processes. In acidic waters the latter are more important, and are due to 'iron bacteria' such as *Gallionella ferruginea* and *Leptothrix ochracea*. The bacteria derive energy from the oxidation. Ferric salts hydrolyze to ferric oxide, which through hydration gives ochre.

High levels of ferrous iron in streams are toxic to some plants and can reduce invertebrate diversity. Ochre precipitate can smother aquatic organisms.

Indicates: Iron precipitation

References: Carlile (2005), Ghiorse (1984), Rasmussen & Lindegaard (1988), Robbins (2006), Snowdon & Wheeler (1993), Wellnitz *et al.* (1994)

'Oily' films on water

These can be either pollution (e.g. a hydrocarbon spill) or a bacterial film (sometimes *Leptothrix discophora*).

Test: Break the film with a stick. If it stays broken, it is a bacterial film. If it flows back into place, it is a hydrocarbon spill.

References: Robbins (2006)

Moist site indicators

Shaded, wall-side verge, Broadway, Worcestershire, 12 May

Dry site indicators

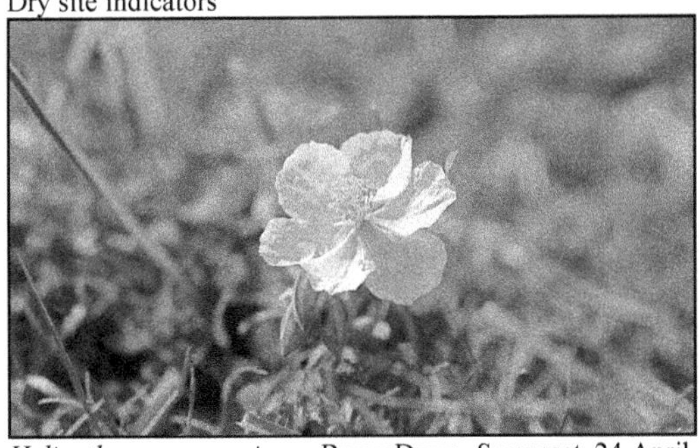

Helianthemum appeninum, Brean Down, Somerset, 24 April

A nitrophilous marsh community

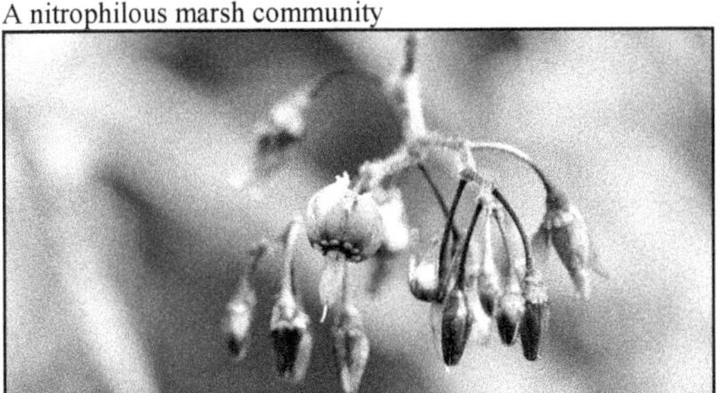

Solanum dulcamara (woody nightshade) Broadway, Worcestershire, 25 July

'Polluted' and 'unpolluted' water

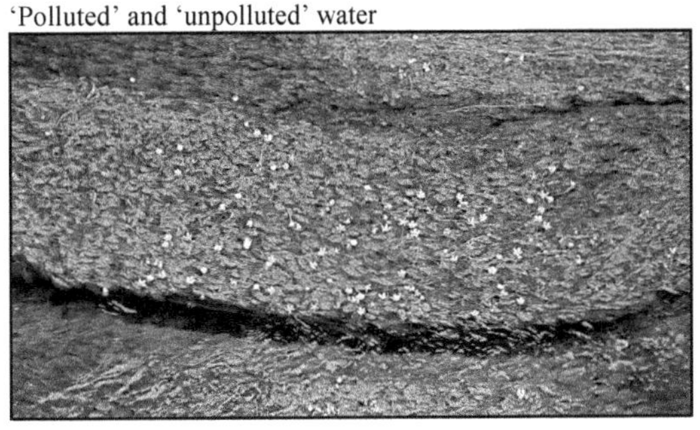

Ranunculus aquatilis (common water-crowfoot) indicates 'clean' water, River Yeo, Somerset, 18 August

Eutrophicated (nutrient-rich) water

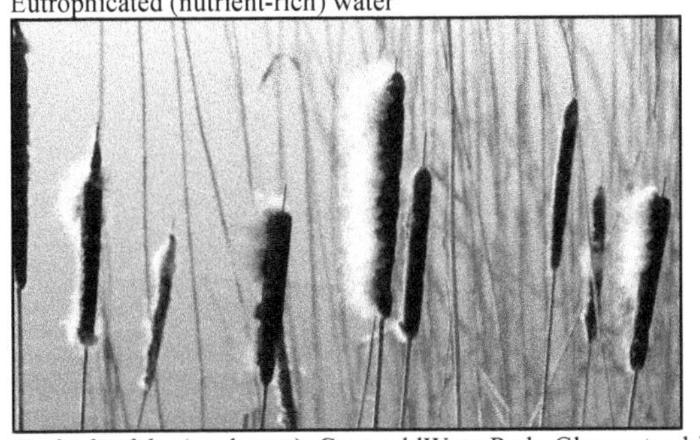

Typha latifolia (reedmace), CotswoldWaterPark, Gloucestershire, 17 February

Oligotrophic (nutrient-poor) water

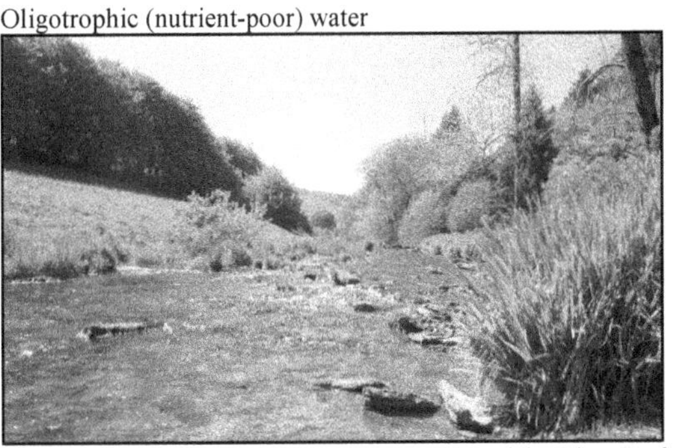

Oligotrophic conditions on the upper reaches of the River Barle (Exmoor), 27 May

Orange precipitates of ochre in streams or springs

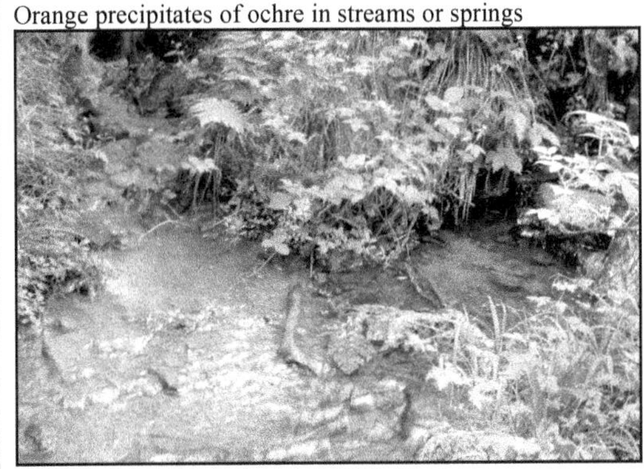

Chalybeate spring, Halesowen, 11 June

'Oily' films on water

Bacterial film on pond, Preston Montford, Shropshire, 30 October

FOAM ON THE SURFACE OF LAKES OR RIVERS

Foam sometimes appears on the surface of streams, rivers and lakes and causes concern. Sometimes it is of natural origin, sometimes synthetic (i.e. pollution) – this can be difficult to determine.

Natural foams can be harmless. Foams caused by pollutants can be harmful.

A foam which persisted for 30 years on the River Rhine caused public concern but correlated with high levels of saponins from water-crowfoot (*Ranunculus fluitans)*, not with synthetic detergents.

Foam is the result of surfactants that reduce the surface tension in water allowing bubbles to form when it is disturbed. Surfactants can be man-made (e.g. detergents) or natural 'dissolved organic carbon' (DOC) arising from the breakdown of algae/plants. Waters with high DOC may be tea coloured (dystrophic). Modern detergents are biodegradable and generally do not produce *persistent* foams but some industrial surfactants are persistent.

Synthetic foam generally accumulates near the source (discharge point). It is white and scented (soapy or fragrant), its appearance will not be related to storms or wind.

Natural foam may:

- occur at a number of locations along a watercourse (e.g. against windward obstacles);
- be white at first but turn brownish as it accumulates sediment;
- persist;
- appear in response to rainstorms, wind, soil erosion events, high temperatures accelerating decay;
- smell of earth, fish or cut grass;

Foaming is more likely in 'soft' waters, just as soap lathers more readily in them.

Test: Shake a sample of spumous water in a jar, natural foam will dissipate, foam from detergents will generally increase.

Caution: These observations enable the gathering of useful data appropriate to later detailed assessment but are no alternative to specialist chemical analysis. Care should be taken to avoid exposure to unknown substances and it is important to remember that 'natural' does not equate with 'harmless'.

References: Wegner & Hamburger (2002), IDEM (2003), Pike (2003)

DEDUCING HEDGEROW AND WOODLAND AGE

Common questions which arise when evaluating the ecological significance of hedgerows and woodlands in the English landscape relate to their ages. Detailed investigations of these features may draw upon evidence from many sources but some simple indicators are tabulated below.

Hedgerows Rule of thumb, 100 years of age for every tree or shrub species present in one 30 m stretch. The rule of thumb should be used with care. Multi-species hedges may have been planted in the past and are planted now. Species counts can exaggerate hedgerow age. Species-rich hedgerows may contain 5 or more native woody species on average in a 30 m length (4 or more in northern England, upland Wales and Scotland). **Indicate: 100 years age per woody species in 30 paces** References: Pollard *et al.* (1974)
Ancient Woodland Indicator Species No plant is an infallible indicator of ancient woodland. The test is whether a wood contains a suite of ancient-woodland plants. Useful indicator species vary across the UK but strong general indicators include: *Melica uniflora* (wood melick), *Galium odoratum* (woodruff), *Anemone nemorosa* (wood anemone), *Luzula pilosa* (hairy wood-rush), *Sorbus torminalis* (wild service-tree), *Melampyrum pratense* (common cow-wheat), *Milium effusum* (wood millet), *Tilia cordata* (small-leaved lime), *Adoxa moschatellina* (moschatel), *Euphorbia amygdaloides* (wood spurge), *Paris quadrifolia* (herb Paris), *Carex strigosa* (thin-spiked wood-sedge), *Carex pallescens* (pale sedge), *Carex pendula* (pendulous sedge), *Carex remota* (remote sedge), *Neottia nidus-avis* (bird's-nest orchid), *Oxalis acetosella* (wood sorrel), *Veronica montana* (wood speedwell), *Allium ursinum* (ramsons), *Convallaria majalis* (lily-of-the-valley), *Lamiastrum galaeobdolon* (yellow archangel). *Galium odoratum* (woodruff) and *Melica uniflora* (wood melick) are particularly good indicators. **Indicate: Ancient woodland indicators** Reference: Rackham (2006)

3. FIELD INDEPENDENT COGNISANCE: SEEING WHAT IS NECESSARY

When interpreting the landscape through ecological features, it is one thing to have open eyes; it is another thing to really see what is important. This is a question of focus.

Good ecological field survey has always been one of the most mentally demanding exercises that the ecological practitioner undertakes, but the parameters are becoming more severe year by year as more species come under threat. The discipline (a word used deliberately) requires good powers of observation and the ability to pick out the important elements from the background ecological noise, e.g. to spot the one specimen of red data book species in the field among all the commoner taxa. Some people have a stronger capacity than others to notice an object that is slightly discordant in, or different to, its surroundings, even in a complex environment. Such people are referred to in studies of cognitive style as '**field independent**'.

Herman Witkin's **field dependence - independence model** identifies an individual's perceptive behaviour while distinguishing object figures from the content field in which they are set (e.g. Witkin *et al.*, 1977). Among other things, field independent learners are able to break up an organized visual field and keep part of it separate; they perceive analytically. These attributes are well suited to good field biology when the ability to sift through large amounts of sensory data (bird song, plant shapes, soil texture/colour, animal scents), discarding background and identifying what is important, in a short amount of time, is a prerequisite for a successful relevant survey. In essence the issue is one of 'focus', a state of mind which can be developed through practice.

Focus and observation can be practised and thus improved. In classical botanical and zoological degrees, undergraduate students were required to spend many hours in a year drawing specimens of animals and plants from life, with hand lenses and through the microscope. The purpose of this, aside from gaining familiarity with the organisms concerned, was to learn to focus and observe critical detail. Even photomicrography does not entrain the same skills, though it can benefit considerably from their application. Without such focus, surveys become superficial and omissions are made. At the end of a day's field survey, it may take the surveyor an hour or two to widen perspective again. After three days of intensive botanical survey, it can take a couple of days before the surveyor stops mentally noting target plant forms as that individual walks about performing other tasks.

In essence, the consultant ecological surveyor or field naturalist should look to the skills and traditions of the Victorian naturalists who accompanied the sailing ships on their scientific journeys across the world. In the absence of cameras, these forerunners of the modern consultant ecologist had to rely on observation, description and drawing (and sometimes the preservation of actual specimens). Considering the journey times involved, they generally had no chance to go back and check on the things they missed and it was therefore important to record everything on the first, and only, visit to the new site.

4. Conclusions

As ecologists and naturalists we have a responsibility to the care of the ecosystems with which we come into contact. In this regard the duty is no less than that of the physician as laid down by Hippocrates over two thousand years ago, i.e.

> *'to help, or at least to do no harm.'*

Hippocrates 400 BCE (Epidemics Book I, Chapter XI) (refer to Smith, 2005)

Accurate field survey and the interpretation of field observation is essential to avoiding ecological harm, especially in the process of ecological impact assessment where the conclusions of the practitioner may lead to significant landuse changes by a developer or land manager. To this end, these *indicators* have been assembled to assist field ecologists in their interpretation of the natural environment through the plants (and certain other species) and the assemblages of such, which they might come across reasonably frequently while surveying in lowland UK. The list cannot be exhaustive as such features can be assembled and read in many different ways just like the letters in an alphabet and there are far more indicators than have been dealt with here. It is the author's hope however that this short introduction will enable the reader to gain a better understanding of the use of such indicators in ecological deduction and will at least assist professional ecologists in living up to Hippocrates' maxim with regard to their interactions with the natural world.

It is intended that this guide should be insightful. A definitive understanding of the effective and practical use of indicator plants can only be achieved through thorough testing of the deductions made. With this thought in mind, the tables are presented as the author's current understanding, and comment on the usefulness or otherwise of the indicators proposed is welcomed.

5. Glossary

Achlorophyllous: Green plants contain chlorophyll which enables them to use energy from sunlight to manufacture carbohydrates from carbon dioxide and water, a process known as photosynthesis. Achlorophyllous plants have no chlorophyll, show no green colouration and cannot use sunlight in this way. They are therefore dependent on other means of carbon nutrition such as parasitism.

Anthropogenic: Generated by human beings. A road verge is an example of an anthropogenic habitat, trampling by people is an example of an anthropogenic disturbance.

Assemblage: In the botanical sense, agroup of different species of plants.

Axiophytes: The concept of axiophytes or 'worthy plants' has been developed by the Botanical Society of the British Isles (BSBI). They are species that are considered to be indicators of habitats important for conservation. For more details refer to the BSBI website (www.bsbi.org.uk/axiophytes.htm).

BAP: Biodiversity Action Plan. A BAP is an internationally recognised programme intended to address the protection of threatened species and habitats. BAPs are produced in response to the International Convention on Biological Diversity signed by world leaders in Rio de Janeiro in 1992.

Base: In simple terms, a base is the opposite of an acid and will give an alkaline reaction when dissolved in water. Calcium carbonate (chalk) is a base and soils derived from chalk or limestone therefore tend to be alkaline in nature. Dolomitic limestone is formed from calcium magnesium carbonate which also gives a base reaction.

Bryophyte: A moss, liverwort or hornwort.

BSBI: Botanical Society of the British Isles.

Calaminarian grasslands: A type of open grassland vegetation that develops on heavy-metal rich soils derived from mine spoil or natural metal-bearing deposits. The word 'calaminarian' relates to calamine, a zinc ore that lends its name to calamine lotion.

Calcareous: Used when referring to soils that contain a large amount of calcium carbonate, often derived from underlying chalk or limestone bedrock.

Calcicole: A plant that thrives on calcareous soils (see 'calcareous') and cannot tolerate acidic soil conditions.

Calcifuge: A plant that cannot tolerate calcareous or other types of base-rich soil.

Chalybeate: Used to describe water containing iron.

Crustose: Term used to describe those lichen species that form a crust on the surface where they grow. The crust is closely attached to the surface with no free edges. It is *not* possible to insert a fingernail beneath the edge of the lichen and peel it away from the substrate (c.f. foliose).

DOC: Dissolved organic carbon. DOC includes many different types of organic molecules found in natural waters that are derived from plants, animals and soils. Waters with a high concentration of DOC are often tea-coloured. Such waters are referred to as dystrophic.

DoE: Department of the Environment (now Department for Environment, Food and Rural Affairs, Defra).

Dystrophic: See DOC above.

Ectomycorrhizal: An adjective describing a type of mycorrhiza where the fungal partner forms a sheath of fungal tissue over the outside of the participating plant root. Many forest trees are ectomycorrhizal.

Elaiosome: Fat- or oil-containing structures formed on the seeds of many plant species that may encourage their collection by foraging ants. The plant benefits because having used the elaiosome for food, the ants discard the seeds in their nest, effectively having planted them in an environment suitable for germination.

Ellenberg's indicator values: Ellenberg (1979, 1988) and Ellenberg *et al.* (1991) devised a set of values which allow the use of European plants as indicators of key environmental factors. The values are based upon individual plant species' known tolerances to environmental parameters such as light, moisture and soil acidity. Among other ecological applications, these indicator values may beof use in the detection of long-term ecological change.

Endemic: A species that occurs nowhere else in the world is endemic to a particular geographic region, e.g. *Fumaria occidentalis* (western ramping fumitory) is endemic to England. It is thought to occur nowhere else in the wild.

Epiparasitic: Literally to be a parasite of a parasite. An epiparasite may be a parasite of another organism via an intermediary. Among the orchids, *Neottia nidus-avis* (bird's-nest orchid) is able to take nutrients from tree roots by parasitizing the mycorrhizal fungi attached to them. It is therefore an epiparasite, having only an indirect connection with the tree root through the intermediary of the fungus but nevertheless benefiting from the tree's nutrient supply (a mycoheterotroph).

Epiphytic: A plant which grows on the surface of another plant without being involved in a parasitic relationship with it, e.g. a fern growing on the branch of a tree.

Eutrophic: A water body with high or excessive nutrient levels and high productivity (c.f. mesotrophic and oligotrophic).

Flocculate: Small suspended particles aggregated to form bigger ones (noun); the aggregation of small suspended particles into bigger ones (verb).

Foliose: Term used to describe those lichen species that are flattened and leaf-like. It is possible to insert a fingernail beneath the edge of the lichen and peel it away from the substrate (c.f. crustose).

Fruticose: Term used to describe those lichen species that are constructed with a radial symmetry and usually attached to the substrate by a single point. The form is therefore various (e.g. goblet-shaped, branched, horn-like) but not leaf-like.

Halophyte: Plants able to tolerate high levels of salt in their environment.

Hyperaccumulate: The characteristic of those plants that accumulate abnormally high levels of certain substances such as arsenic or zinc in their tissues. Hyperaccumulators are able to tolerate levels of these substances which would be toxic or lethal to other organisms.

Katharobic: Term used to describe extremely pure natural waters where organic matter is at such a minimum that any aquatic flora is limited e.g. some spring waters.

Macrophyte: Term used to describe aquatic plants that are not microscopic.

Macroscopic: Visible to the naked eye.

Mesotrophic:A water body with moderate nutrient levels and productivity (c.f. eutrophic and oligotrophic).

Metalliferous: Metal-bearing.

Metallophyte:A plant that can tolerate high levels of heavy metal in its environment. Obligate metallophytes are dependent upon the presence of such levels.

Monomorphic:A monomorphic population is one consisting of individuals with all the same trait (or traits) where the trait can exist in different forms. A polymorphic population consists of individuals where the trait varies between individuals e.g. blue and brown eye colour in the same species.

Mycoheterotrophic: Term describing achlorophyllous plants that obtain their carbon requirements through symbiotic associations with fungi (c.f. achlorophyllous) (refer to Leake, 2004).

Mycorrhizal: An adjective referring to plants and soil fungi which enter into symbiotic relationships with each other leading to the development of mycorrhizae (singular mycorrhiza), structures formed of fungal hyphae (threads) and plant root cells from the organisms involved. The participating fungus and plant exchange nutrients through the mycorrhiza to their mutual benefit. Most land plants are mycorrhizal.

Nitrophile:A plant that thrives on soils rich in available forms of nitrogen e.g. nitrate and ammonium compounds (a nitrophilous plant). Nitrogen itself is a gas and cannot be used directly by plants although it can be used by some bacteria (some of which exist in symbiotic association with certain plant species).

Nitrophobe: Used in the sense of Pitcairn *et al.* (2006) to mean a non-nitrogen-loving plant, i.e. one that is sensitive to high levels of available nitrogen deposition and therefore may be adversely affected in some way.

NVC: National Vegetation Classification. The NVC is a phytosociological classification of the terrestrial, freshwater and maritime plant communities found in Great Britain (not includingNorthern Ireland).

Oligotrophic:A water body with low nutrient levels and low productivity (c.f. eutrophic and mesotrophic).

Rhizome: An underground or ground-level stem which is usually horizontal with buds and scale leaves.

Ruderal:A plant or vegetation of wasteland and fallow or disturbed places.

Serpentine: The word serpentine is used broadly to describe the rock, minerals, soil and vegetation associated with serpentinized, ultrabasic rocks. These all contain magnesium and iron as ferromagnesian silicates; other common elements include nickel, chromium, and cobalt. Soils derived from them are toxic to many plant species and support unusual floras (refer to Kruckeberg 2002).

Siliceous:A term used to describe silica-bearing rock types.

Symbiotic: Relating to symbiosis i.e. two dissimilar organisms living together in close association. Sometimes used synonymously with mutualism where the relationship is beneficial to both organisms involved.

Thallus (plural thalli):A relatively simple plant body characteristic of liverworts, lichens, algae and fungi that is not differentiated into stems, roots and leaves.

Urban heat island: A phenomenon resulting in warmer conditions prevailing in cities and other urban areas compared with the surrounding countryside, a result of activities such as domestic and industrial heating, lower evapotranspiration in the built environment and the absorption of heat by artificial surfaces.

6. REFERENCES

Amphlett, J. & Rea, C. (1909). *The Botany of Worcestershire.*Birmingham, UK: Cornish Brothers Ltd.

Arnolds, E. (1981). *Ecology and Coenology of Macrofungi in Grasslands, in Drenthe, The Netherlands: Introduction and Synecology.* Vol. 1 (Part 1). Vaduz, Germany: Bibliotheca Mycologica, Gantner Verlag K.G.

Arnolds, E. (1982). *Ecology and Coenology of Macrofungi in Grasslands, in Drenthe, The Netherlands: Autecology and Taxonomy.* Vol. 2 (Parts 2 and 3). Vaduz, Germany: Bibliotheca Mycologica, Gantner Verlag K.G.

Atherton, I., Bosanquet, S. & Lawley, M. (2010) *Mosses and Liverworts of Britain and Ireland: A Field Guide.* British Bryological Society.

Atkinson, M. (1996). The distribution and naturalisation of *Lathraea clandestina* L. (Orobanchaceae) in the British Isles. *Watsonia, 21,* 119–128.

Atkinson, M. & Atkinson, E. (2002). Biological flora of the British Isles: *Sambucus nigra* L.*Journal of Ecology, 90,* 895–923.

Baker, A. & Brooks, R. (1988). Botanical exploration for minerals in the humid tropics.*Journal of Biogeography, 15,* 221–229.

Baker, A. & Dalby, D. (1980). Morphological variation between some isolated populations of *Silene maritima* within the British Isles with particular reference to inland populations on metalliferous soils. *New Phytologist, 84,* 123–133.

Bannister, P. (1966). Biological flora of the British Isles: *Erica tetralix* L.*Journal of Ecology, 54,* 795–813.

Barnes, G. & Williamson, T. (2006). *Hedgerow History: Ecology, History and Landscape Character.*Macclesfield, UK: Windgather Press.

Beckett, G. & Bull, A. (1999). *A Flora of Norfolk.*Thetford, UK: Privately printed.

Bedfordshire Wildlife Trust. (2009). *St Andrew's Churchyard*. Biological Report, Recorder: Laura Downton.

Björkman, E. (1960). *Monotropa hypopitys* L. - an epiparasite on tree roots. *Physiologia Plantarum, 13,* 308–327.

Blackman, G. & Rutter, A. (1954). Biological flora of the British Isles: *Endymion (Hyacinthoides) non-scriptus* (L.) Garcke (*non-scripta* (L.) Chouard ex Rothm.).*Journal of Ecology, 42,* 629–638.

Bobbink, R. & Lamers, L. (2002). Effects of increased nitrogen deposition. In J. Bell& M. Treshow (Eds.), *Air Pollution and Plant Life.* 2nd Edn. Chichester: John Wiley and Sons Ltd.

Bold, H. & Wynne, M. (1978). *Introduction to the Algae: Structure and Reproduction.*Englewood Cliffs, New Jersey, USA: Prentice-Hall, Inc.

ter Borg, S. (1972). *Variability of*Rhinanthus serotinus *(Schonh.) Obory in Relation to the Environment.* PhD thesis, University of Gröningen, Germany.

Bowen, H. (2000). Thorow-wax as a florist's alien. *BSBI News85,* 46.

Box, J., Brown, M., Coppin, N., Hawkeswood, N., Webb, M., Hill, A. *et al.* (2011). Experimental wet heath translocation in Dorset, England. *Ecological Engineering, 37,* 158–171.

Brunet, J. & von Oheimb, G. (1998). Migration of vascular plants to secondary woodlands in southern Sweden. *Journal of Ecology, 86,* 429–438.

Brys, R. & Jacquemyn, H. (2009). Biological flora of the British Isles: *Primula veris.**Journal of Ecology, 97,* 581–600.

Burdon, J. (1983). Biological flora of the British Isles: *Trifolium repens* L. *Journal of Ecology, 71*, 307–330.

Carlile, M. (2005). The red streams of east Berkshire. In M. Crawley (Ed.), *The Flora of Berkshire.* Harpenden, UK: Brambleby Books.

Castroviejo, S., Laínz, M., Lopez González, G., Monserrat, P., Muñoz Gormendia, F., Paiva, J. *et al.* (1986). *Flora Iberica: Plantas Vasculares de la Península Ibérica e Islas Baleares.* Vol. 1: Lycopodiaceae – Papaveraceae. Madrid, Spain: Real Jardín Botánico, Consejo Superior de Investigaciones Científicas.

Clapham, A., Tutin, T. & Warburg, E. (1962). *Flora of the British Isles.* Cambridge, UK: CambridgeUniversity Press.

Clark, C., Cleland, E., Collins, S., Fargione, J., Gough, L., Gross, K., *et al.* (2007) Environmental and plant community determinants of species loss following nitrogen enrichment. *Ecology Letters*, 10, 596-607.

Clausen, R. (1938). A Monograph of the Ophioglossaceae. *Memoirs of the Torrey Botanical Club, 19*, 3–173.

Cleland, E. & Harpole, S. (2010) Nitrogen enrichment and plant communities. *Annals of the New YorkAcademy of Sciences,* 1195, 46-61.

Clement, E. & Foster, M. (1994). *Alien Plants of the British Isles.* London, UK: Botanical Society of the British Isles.

Compton, J. & Boone, R. (2000). Long-term impacts of agriculture on soil carbon and nitrogen in New England forests. *Ecology, 81*, 2314–2330.

Conan Doyle, A. (1892). The Adventure of the Blue Carbuncle. *Strand Magazine.*

Crawley, M. (2005). *The Flora of Berkshire.* Harpenden, UK: Brambleby Books.

Crowther, K., Bliss, A. & Smith, P. (2009). Adder's-tongue: a 13-year translocation story. *In Practice64*, 18–22.

Cubero, R. & Moreno, M. (1979). Agronomic control and sources of resistance in *Vicia faba* to *Orobanche* spp. In G. Artiscia-Mugnozza & M. Poulsen (Eds.), *Some Current Research on* Vicia faba *in Western Europe* (pp. 41-80). Luxemburg: Commission of the European Communities.

Daniel, P., Woodward, F., Bryant, J. & Etherington, J. (1985). Nocturnal accumulation of acid in leaves of wall pennywort (*Umbilicus rupestris*) following exposure to water stress. *Annals of Botany, 55*, 217–223.

Davies, D. & Graves, J. (2000). The impact of phosphorus on interactions of the hemiparasitic angiosperm *Rhinanthus minor* and its host *Lolium perenne. Oecologia,124*, 100–106.

Davison, A. (1971). The ecology of *Hordeum murinum* L.: II the ruderal habit. *Journal of Ecology, 59*, 493–506.

Davy, A. (1980). Biological flora of the British Isles: *Deschampsia caespitosa. Journal of Ecology, 68*, 1075–1096.

Defra (2004, revised 2007). *Code of Practice on How to Prevent the Spread of Ragwort.* Crown copyright. UK: Defra.

Department of the Environment (DoE). (1994). *Contaminated Land Research Report: Guidance on Preliminary Site Inspection of Contaminated Land.* Prepared by: Applied Environmental Research Centre Ltd. Department of the Environment.

Dobson, F. (2005). *Lichens: An Illustrated Guide to the British and Irish Species.* 5th Edn. Slough, UK: The Richmond Publishing Co. Ltd.

Dodds, W., Gudder, D. & Mollenhauer, D. (1995). The ecology of *Nostoc. Journal of Phycology, 31*, 2–18.

Dowdeswell, W. (1987). *Hedgerows and Verges.* London, UK: Allen and Unwin.

Edgington, J. (2002). *Umbilicus rupestris* - Eastward Ho? *BSBI News90*, 12–13.

Eidt, R. (1977). Detection and examination of anthrosols by phosphate analysis. *Science, 1977,* 1327–1333.

Ellenberg, H. (1979). Zeigerwerte der Gefasspflanzen Mitteleuropas. *Scripta Geobotanica, 9,* 1–122.

Ellenberg, H. (1988). *Vegetation Ecology of Central Europe.* 4th Edn. Cambridge, UK: CambridgeUniversity Press.

Ellenberg, H., Weber, H., Dull, R., Wirth, V., Werner, W. & Paulißen, D. (1991). Indicator values of plants in central Europe. *Scripta Geobotanica, 18,* 1–248.

Ellis, M. & Ellis, J. (1997) *Microfungi on Land Plants: An Identification Handbook.* 2nd Edn. Slough, UK: The Richmond Publishing Co. Ltd.

Fairbairn, C. & Thomas, B. (1959). The potential nutritive value of some weeds common to north-eastern England. *Journal of the British Grassland Society, 14,* 36–46.

Fish, J. & Fish, S. (2010). *A Student's Guide to the Seashore.* 3rd Edn. Cambridge, UK: CambridgeUniversity Press.

Foley, M. & Clarke, S. (2005). *Orchids of the British Isles.*Cheltenham, UK: Griffin Press Publishing Ltd.

Forsberg, C. (1965). Environmental conditions of Swedish charophytes. *Symbolae Botanicae Upsalienses, 18,* 1–67.

Ghiorse, W. (1984). Biology of iron- and manganese-depositing bacteria. *Annual Review of Microbiology, 38,* 515–550.

Gilbert, O. (1991). *The Ecology of Urban Habitats.*London, UK: Chapman and Hall.

Gilbert, O. (2000). *Lichens.* The New Naturalist. London, UK: HarperCollins Publishers.

Gilbert, O. & Anderson, P. (1998). *Habitat Creation and Repair.*Oxford, UK: OxfordUniversity Press.

Gillam, B. (1993). *The Wiltshire Flora.*Newbury, UK: Pisces Publications.

Griffith, G., Easton, G. & Jones, A. (2002). Ecology and diversity of waxcap (*Hygrocybe* spp.) fungi. *Botanical Journal of Scotland, 54,* 7–22.

Grigson, G. (1958). *The Englishman's Flora.*St Albans, UK: Paladin.

Grime, J. (1963). Factors determining the occurrence of calcifuge species on shallow soils over calcareous substrata. *Journal of Ecology, 51,* 375–390.

Grime, J. (1979). *Plant Strategies and Vegetation Processes.*Chichester, UK: John Wiley & Sons.

Grime, J., Hodgson, J. & Hunt, R. (1990) *The Abridged Comparative Plant Ecology.* London, UK: Unwin Hyman.

Grime, J., Thompson, K., Hunt, R., Hodgson, J., Cornelissen, J., Rorison *et al.* (1997). Integrated screening validates primary axes of specialisation in plants. *Oikos,* 79, 259–281.

Grime, J., Hodgson, J. & Hunt, R. (1988). *Comparative Plant Ecology: A Functional Approach to Common British Species.* 1st Edn. London, UK: Unwin Hyman.

Grime, J., Hodgson, J. & Hunt, R. (2007). *Comparative Plant Ecology: A Functional Approach to Common British Species.* 2nd Edn. Dalbeattie, UK: Castlepoint Press.

Grose, J. (1957). *The Flora of Wiltshire.* Devises, UK: Wiltshire Archaeological and Natural History Society.

Hahn, S., Tenhunen, J., Popp, P. & Lange, O. (1993). Upland tundra in the foothills of the Brooks range, Alaska: diurnal CO_2 exchange patterns of characteristic lichen species. *Flora, 188,* 125–143.

Hakan, R., Jeglum, J. & Hooijer, A. (2006). *The Biology of Peatlands: The Biology of Habitats.*Oxford, UK: OxfordUniversity Press.

Harper, J. (1957). Biological flora of the British Isles: *Ranunculus acris* L., *R. repens* L. and *R. bulbosus* L. *Journal of Ecology, 45*, 289–342.

Haslam, S. (1990). *River Pollution: An Ecological Perspective.*London, UK: Belhaven Press.

Hedderson, T., Letts, J. & Payne, K. (2003a). Bryophyte diversity and community structure on thatched roofs of the Holncote estate, Somerset, UK. *Journal of Bryology, 25*, 49–60.

Hedderson, T., Letts, J. & Payne, K. (2003b). The rare thatch moss, *Leptodontium gemmascens*, on the Holncote estate, Somerset, UK: distribution and abundance in relation to roof variables. *Lindbergia, 28*, 113–119.

Heinricher, E. (1894). Die Keimung von *Lathraea. Berichte der Deutschen Botanischen Gesellschaft, 12*, 117–132.

Hendrickson, W. (1972). Perspective on fire and ecosystems in the United States. *Fire in the Environment: Symposium Proceedings* (pp. 29–33). Denver, USA: USDA Forest Service, in cooperation with Fire Services of Canada, Mexico, and the United States; Members of the Fire Management Study Group; North American Forestry Comission.

Hermy, M., Honnay, O. & Laweson, J. (1999). An ecological comparison between ancient and other forest plant species of Europe, and the implications for forest conservation. *Biological Conservation, 91*, 9–22.

Hill, M., Preston, C. & Smith, A. (1994). *Atlas of the Bryophytes of Britain and Ireland: Mosses (Diplolepidae).* Vol. 3. Colchester, UK: Harley Books.

Hill, M., Mountford, J., Roy, D. & Bunce, R. (1999). *Ellenberg's Indicator Values for British Plants.*Norwich, UK: HMSO.

Hill, M., Preston, C. & Roy, D. (2004). *PLANTATT: Attributes of British and Irish Plants: Status, Size, Life History, Geography and Habitats.*Cambridge, UK: Centre for Ecology and Hydrology.

Hill, M., Preston, C., Bosanquet, S. & Roy, D. (2007). *BRYOATT: Attributes and British and Irish Mosses, Liverworts and Hornworts.*Cambridge, UK: Centre for Ecology and Hydrology.

Hippocrates 400 BCE Epidemics Book I, Chapter XI

Hoffman, G. (1966). Ecological studies of *Funaria hygrometrica* Hedw. in eastern Washington and northern Idaho. *Ecological Monographs, 36*, 157–180.

Honnay, O., Hermy, M. & Coppin, P. (1999). Impact of habitat quality on forest plant species colonization. *Forest Ecology and Management, 115*, 157–170.

Hornby, R. & Rose, F. (1986). *The Use of Vascular Plants in Evaluation of Ancient Woodland for Nature Conservation in Southern England.* Unpublished Nature Conservancy Council Report.

Humphries, C. & Shaughnessy, E. (1987) *Gorse.* Shire Natural History Series No 9. Princes Risborough, UK: Shire Publications Ltd.

Hutchings, M. & Price, E. (1999). Biological flora of the British Isles: *Glechoma hederacea* L. (*Nepeta glechoma* Benth., *N. hederacea* (L.) Trev.). *Journal of Ecology, 87*, 347–364.

Indiana Department of Environmental Management (IDEM) (2003). *Foam Fact Sheet.*Retrieved 2010, 9-June from Indiana Department of Environmental Management: http://www.in.gov /idem/4561.htm.

Jacquemyn, H., Brys, R. & Hutchings, M. (2008). Biological flora of the British Isles: *Paris quadrifolia. Journal of Ecology, 96*, 833–844.

Jacquemyn, H., Brys, R., Honnay, O. & Hutchings, M. (2009). Biological flora of the British Isles: *Orchis mascula* (L.) L. *Journal of Ecology, 97*, 360–377.

Joint Nature Conservation Committee (JNCC). (1993). *Handbook for Phase 1 Habitat Survey: A Technique for Environmental Audit* (Revised 2003). Peterborough, UK: Nature Conservancy Council.

Josselyn, J. (1672). *New England's Rarities.* Reproduced with an introduction by E. Tuckerman (1865). Boston, USA: William Veazie.

Kay, Q. (1994). The history, ecology and distribution of the flora of Glamorgan. In A. Wade, Q. Kay, R. Ellis & N. Wales (Eds.), *Flora of Glamorgan.*London, UK: HMSO.

Kershaw, G. & Kershaw, L. (1986). Ecological characteristics of 35-year-old crude-oil spills in tundra plant communities of the Mackenzie mountains, N.W.T. *Canadian Journal of Botany, 64,* 2935-2947.

King, T. (1981). Ant-hills and grassland history. *Journal of Biogeography, 8,* 329–334.

Klinkhamer, P. & de Jong, T. (1993). Biological flora of the British Isles: *Cirsium vulgare* (Savi) Ten. *Journal of Ecology, 81,* 177–191.

Knight, G. (1964). Some factors affecting the distribution of *Endymion non-scriptus* (L.) Garcke in Warwickshire woods. *Journal of Ecology, 52,* 405–421.

Kowarik, I. & Säumel, I. (2007) Biological flora of Central Europe: *Ailanthus altissima* (Mill.) Swingle. *Perspectives in Plant Ecology, Evolution and Systematics 8,* 207–237.

Kretzschmar, H., Eccarius, W. & Dietrich, H. (2007). *The Orchid Genera* Anacamptis, Orchis *and* Neotinea*: Phylogeny, Taxonomy, Morphology, Biology, Distribution, Ecology and Hybridisation.*Burgel, Germany: Echinomedia Verlag.

Kruckeberg, A. (2002). *Geology and Plant Life: The Effects of Landforms and Rock Types on Plants.*Seattle, USA: University of Washington Press.

Kuijt, J. (1969). *The Biology of Parasitic Flowering Plants.*Berkeley and Los Angeles, USA: University of California Press.

Lancaster, R. (1990). *Lathraea clandestina. The Garden, 115,* 145–147.

Leake, J. (2004). Myco-heterotroph/epiparasitic plant interactions with ectomycorrhizal and arbuscular mycorrhizal fungi. *Current Opinion in Plant Biology, 7,* 422–428.

Leith, I. & Fowler, D. (1987) Urban distribution of *Rhytisma acerinum* (Pers.) Fries (Tar Spot) on Sycamore. *New Phytologist,* 108, 175-181.

Lepp, H. (2007). *Australian Bryophytes - Case Studies: Ceratodon purpureus.* Retrieved 09 June 2010 from http://www.anbg.gov.au/bryophyte/case-studies/fire-ceratodon-purpureus.html.

Lepp, H. (2008). *Bryophyte Ecology: Fire.* Retrieved 08 June 2010 from http://www.anbg.gov.au/bryophyte/ecology-fire.html.

Lewandowski, A. (2000). Compaction. *The Soil Management Series* 2.

Linke, K.-H., Sauerborn, J. & Saxena, M. (1989). *Orobanche Field Guide.*Aleppo, Syria: ICARDA.

Lopez-Granados, F. & Garcia-Torres, L. (1999). Longevity of crenate broomrape (*Orobanche crenata*) seed under soil and laboratory conditions. *Weed Science, 47,* 161–166.

Lousley, J. (1969). *Wild Flowers of the Chalk and Limestone.* The New Naturalist. London, UK: Collins.

Mabey, R. (1996). *Flora Britannica.*London, UK: Sinclair-Stevenson.

Marren, P. (1990). *Britain's Ancient Woodland: Woodland Heritage.*London, UK: David and Charles.

Marrs, R. & Watt, A. (2009). Biological flora of the British Isles: *Pteridium aquilinum* (L.) Kuhn. *Journal of Ecology, 94,* 1272–1321.

Martin, M. (1968). Conditions affecting the distribution of *Mercurialis perennis* L. in certain Cambridgeshire woodlands. *Journal of Ecology, 56,* 777–793.

Martin, M. & Fawcett, K. (1997). The biological implications of heavy metals in the Mendips. *Proceedings of the Bristol Naturalist's Society, 55*, 95–112.

McKendrick, S., Leake, J., Taylor, D. & Read, D. (2002). Symbiotic germination and development of the mycoheterotrophic orchid *Neottia nidus-avis* in nature and its requirement for locally distributed *Sebacina* spp. *New Phytologist, 154*, 233–247.

McLauchlan, K. (2006). The nature and longevity of agricultrual impacts on soil carbon and nutrients: a review. *Ecosystems, 9*, 1364–1382.

Meulebrouck, K. (2009). Distribution, demography and metapopulation dynamics of *Cuscuta epithymum* in managed heathlands. PhD thesis, Katholieke Universiteit Leuven, Belgium.

Miller, H., Clarkson, B. & Smith, P. (2007). *A Strategic Conservation Assessment of Heathland and Associated Habitats on the Coal Spoils of South Wales.* Countryside Council for Wales. Countryside Council for Wales Science Report 772.

Möller, O. (1987). Von Sammenkorn bis zur ersten Knolle: Das Protocormstadium von *Orchis mascula. Di Orchidee, 38*, 297–302.

Moore, J. (1986). *BSBI Handbook No. 5: Charophytes of Great Britain and Ireland.*London, UK: Botanical Society of the British Isles.

Mukerji, S. (1936). Contributions to the autecology of *Mercurialis perennis* L. Parts I-III. *Journal of Ecology, 24*, 38–81.

Muller, S. (2000). Assessing occurrence and habitat of *Ophioglossum vulgatum* L. and other Ophioglossaceae in European forests. Significance for nature conservation. *Biodiversity and Conservation, 9*, 673-681.

Page, C. (1988). *Ferns: Their Habitats in the British and Irish Landscape.* The New Naturalist. London, UK: Collins.

Palmqvist, K. & Sundberg, B. (2000). Light use efficiency of dry matter gain in five macrolichens: relative impact of microclimate conditions and species-specific traits. *Plant Cell Environment, 23*, 1–14.

Parker, C. & Riches, C. (1993). *Parasitic Weeds of the World.* Wallingford, UK: CAB International.

Pavek, D. (1992). *Chamerion angustifolium.* (F. S. U.S. Department of Agriculture, Fire Effects Information System OnLine). Retrieved 08 Feb 2010 from: http://www.fs.fed.us/database/feis/.

Pegler, D., Laessoe, T. & Spooner, B. (1995). *British Puffballs, Earthstars and Stinkhorns: An Account of the British Gasteroid Fungi.*Kew, UK: Royal Botanic Gardens.

Peterken, G. (1974). A method for assessing woodland flora for conservation using indicator species. *Biological Conservation, 6*, 239–245.

Peterken, G. (1981). *Woodland Management and Conservation.*London, UK: Chapman and Hall.

Peterken, G. & Game, M. (1984). Historical factors affecting the number and distribution of vascular plant species in the woodlands of central Lincolnshire. *Journal of Ecology, 72*, 155–182.

Phillips, R. (2006). *Mushrooms.*London, UK: Macmillan.

Pike, R. (2003). Foam: An Introduction. *Streamline Watershed Management Bulletin, 7*, 18.

Pitcairn, C., Leith, I., Sheppard, L. & Sutton, M. (2006). *Development of a Nitrophobe/nitrophile Classification for Woodlands, Grasslands and Upland Vegetation in Scotland.* NERC/Centre for Ecology and Hydrology, 24pp. (CEH: Project Report Number: C03066).

Poel, L. (1948). Effects of aeration on bracken and heather grown in solution. *Nature, 162*, 115.

Poel, L. (1951). Soil aeration in relation to *Pteridium aquilinum* (L.) Kuhn. *Journal of Ecology, 39*, 182–191.

Poel, L. (1961). Soil aeration as a limiting factor in the growth of *Pteridium aquilinum* (L.) Kuhn. *Journal of Ecology, 49*, 107–111.

Pollard, E., Hooper, M. & Moore, N. (1974). *Hedges.* The New Naturalist. London, UK: Collins.

Porley, R. & Hodgetts, N. (2005). *Mosses and Liverworts.* The New Naturalist. London, UK: Collins.

Preston, C. & Croft, J. (1997). *Aquatic Plants in Britain and Ireland.*Colchester, UK: Harley Books.

Preston, R., Pearman, D. & Dines, T. (2002). *New Atlas of the British and Irish Flora.*Oxford, UK: OxfordUniversity Press.

Prime, C. (1981). *Lords and Ladies.* The New Naturalist. London, UK: Collins.

Proctor, J. & Woodell, R. (1971). The plant ecology of serpentine: I. Serpentine vegetation of England and Scotland. *Journal of Ecology, 59*, 375–395.

Rackham, O. (1986). *The History of the Countryside.*London, UK: J. M. Dent.

Rackham, O. (2006). *Woodlands.* The New Naturalist. London, UK: Collins.

Ramsbottom, J. (1953). *Mushrooms and Toadstools: A Study of the Activities of Fungi.* The New Naturalist. London, UK: Collins.

Rasmussen, H. (1995). *Terrestrial Orchids from Seed to Mycotrophic Plant.*Cambridge, UK: CambridgeUniversity Press.

Rasmussen, K. & Lindegaard, C. (1988). Effects of iron compounds on macroinvertebrate communities in a Danish lowland river system. *Water Research, 22*, 1101–1108.

Rayner, M. (1921). The ecology of *Calluna vulgaris*, II. The calcifuge habit. *Journal of Ecology, 9*, 60–74.

Reed, R. & Russell, G. (1978). Salinity fluctuations and their influence on 'bottle brush' morphogenesis in *Enteromorpha intestinalis* (L.) Link. *European Journal of Phycology, 13*, 149–153.

Rich, T. (1991). *Crucifers of Great Britain and Ireland.*London, UK: Botanical Society of the British Isles.

Rich, T. (2001). Flowering plants. In D. Hawksworth (Ed.), *The Changing Wildlife of Britain and Ireland.*London, UK: Taylor and Francis.

Rich, T. & Jermy, A. (1998). *Plant Crib 1998.*London, UK: Botanical Society of the British Isles.

Richards, A., Lefebvre, C., Macklin, M., Nicholson, A. & Vekemans, X. (1989). The population genetics of *Armeria maritima* (Mill.) Willd. on the River South Tyne, UK. *New Phytologist, 112*, 281–293.

Riddelsdell, H., Hedley, G. & Price, W. (1948). *Flora of Gloucestershire: Phanerogams, Vascular Cryptogams, Charophyta.*Bristol, UK: The Chatford House Press Ltd for the Cotteswold Naturalist's Field Club.

Robbins, N. (2006). *How to Collect and See the Microbial Community that Fixes Iron and Managanese in the Natural Environment (For Amateurs and Professionals).* USGS Science for a Changing World. Retrieved 09 June 2010 from http://pubs.usgs.gov/gip/microbes.

Robinson, B., Leblanc, M., Petit, D., Brooks, R., Kirkman, J. & Gregg, P. (1998). The potential of *Thlaspi caerulescens* for phytoremediation of contaminated soils. *Plant and Soil, 203*, 47-56.

Rodwell, J. (1991a). *British Plant Communities Vol. 1: Woodlands and Scrub.* Cambridge, UK: CambridgeUniversity Press.

Rodwell, J. (1991b). *British Plant Communities Vol. 2: Mires and Heaths.* Cambridge, UK: CambridgeUniversity Press.

Rodwell, J. (1992). *British Plant Communities Vol. 3:Grasslands and Montane Communities.* Cambridge, UK: CambridgeUniversity Press.

Rodwell, J. (1995). *British Plant Communities Vol. 4:Aquatic Communities, Swamps and Tall-herb Fens.* Cambridge, UK: CambridgeUniversity Press.

Rodwell, J. (2000). *British Plant Communities Vol. 5:Maritime Communities and Vegetation of Open Habitats.* Cambridge, UK: CambridgeUniversity Press.

Rose, F. (1999). Indicators of ancient woodland: the use of vascular plants in evaluating ancient woods for nature conservation. *British Wildlife,* 241–251.

Round, F. (1973). *The Biology of the Algae.* 2nd Edn. London, UK: Edward Arnold (Publishers) Ltd.

Rural Development Service. (2005). *Higher Level Stewardship: Farm Environment Plan: Guidance Handbook.* Department for Environment, Food and Rural Affairs.

Salisbury, E. (1942). *Reproductive Capacity of Plants.*London, UK: Bell and Sons.

Salisbury, E. (1961). *Weeds and Aliens.* The New Naturalist. London, UK: Collins.

Sandor, J. & Eash, N. (1995). Ancient agricultural soils in the Andes of southern Peru. *Soil Science Society of America Journal, 59,* 170–179.

Schaminée J., Stortelder A. & Weeda E. (1996). De vegetatie van Nederland: deel 3. Plantengemeenschappen van graslanden, zomen en droge heiden. Leiden, Netherlands: Opulus Press.

Schell, D. & Alexander, V. (1973). Nitrogen fixation in Arctic coastal tundra in relation to vegetation and micro-relief. *Arctic, 26,* 130–137.

Schotsman, H. (1972). Note sur la repartition des Callitriches en Sologne et dans les regions limitrophes. *Bulletin du Centre d'Etudes et de Recherches Scientifiques, Biaritz, 9,* 19–52.

Scott, N. (1985). The updated distribution of maritime species on British roadsides. *Watsonia, 15,* 381–386.

Scott, N. & Davison, A. (1982). De-icing salt and the invasion of road verges by maritime plants. *Watsonia, 14,* 41–52.

Seaward, M. (1996). *Lichen Atlas of the British Isles: Fascicle 2 Cladonia Part 1 (59 species).*London, UK: British Lichen Society.

Sellars, B. & Baker, A. (1988). *Review of Metallophyte Vegetation and its Conservation.* CSD report no. 797. Peterborough, UK: Nature Conservancy Council.

Selosse, M., Weiss, M., Jany, J. & Tillier, A. (2002). Communities and populations of sebacinoid basidiomycetes associated with the achlorophyllous orchid *Neottia nidus-avis* (L.) LCM Rich. and neighbouring tree ectomycorrhizae. *Molecular Ecology, 11,* 1831–1844.

Shaw, P. & Kibby, G. (2001). Aliens in the flowerbeds: the fungal diversity of ornamental woodchips. *Field Mycology, 2,* 6–11.

Shaw, P., Butlin, J. & Kibby, G. (2004). Fungi of ornamental woodchips in Surrey. *Mycologist, 18,* 12–15.

Sheppard, A. (1991). Biological flora of the British Isles: *Heracleum sphondylium* L. *Journal of Ecology, 79,* 235–258.

Simkin, J. (2007). *Calaminarian Grasslands in Northern England.* Institute of Ecology and Environmental Management. Retrieved 12 Jan 2010 from www.ieem.net/docs/08%20Janet%Simpkin.pdf.

Sinker, C., Packham, J., Trueman, I., Oswald, P., Perring, F. & Westwood, W. (1991). *Ecological Flora of the Shropshire Region.*Shrewsbury, UK: Shropshire Wildlife Trust.

Smith, P. (1984). *Systematic Studies of British Fumaria Species.* PhD thesis, University of Bristol, UK.

Smith, A. (2001). Mosses, liverworts and hornworts. In D. Hawksworth (Ed.), *The Changing Wildlife of Britain and Ireland* (pp. 78-102). London, UK: Taylor and Francis.

Smith, C. (2005) Origin and uses of *primum non nocere* – above all, do no harm! *Journal of Clinical Pharmacology*, 45, 371-377.

Smith, P. & Grenfell, A. (1984). The genus *Fumaria* in Avon. *Proceedings of the Bristol Naturalists Society, 44,* 35–42.

Smith, P. & Sangwine, T. (2002). Highways: the ecological resource net. *Proceedings of the European Transport Conference 9–11 September 2002.* European Transport Conference, Cambridge, UK.

Smith, C., Aptroot, A., Coppins, B., Fletcher, A., Gilbert, O., James, P. *et al.* (2009). *The Lichens of Great Britain and Ireland.*London, UK: British Lichen Society.

Snowdon, R. & Wheeler, B. (1993). Iron toxicity to fen plant species. *Journal of Ecology, 81,* 35–46.

Söderström, L. (1992). Invasions and range expansions of bryophytes. In J. Bates & A. Farmer (Eds.), *Bryophytes and Lichens in a Changing Environment.*Oxford, UK: Clarendon Press.

Stace, C. (2010). *New Flora of the British Isles.* 3rd Edn. Cambridge, UK: CambridgeUniversity Press.

Stana, D. & Vârban, R. (2005). Biological, ecological and agricultural indicators of grassland flora from Aghiresu area, Cluj District. *Notulae Botanicae Horti Agrobotanici Cluj-Napoca, 33,* 7–14.

Steele, B. (1955). Soil pH and base status as factors in the distribution of calcicoles. *Journal of Ecology, 43,* 120–132.

Stickney, P. (1986). *First Decade Plant Succession Following the SundanceForest Fire, Northern Idaho.* General Technical Report INT-197, U.S. Department of Agriculture, Forest Service, Intermountain Research Station, Ogden, UT.

Stickney, P. (1990). Early development of vegetation following holocaustic fire in Northern Rocky Mountains. *Northwest Science, 64,* 243–246.

Sukkop, H. & Weiler, S. (1986). Biotype mapping in urban areas of the Federal Republic of Germany. *Landschaft und Stadt, 18,* 25–28.

Summerhayes, V. (1968). *Wild Orchids of Britain.* 2nd Edn. The New Naturalist. London, UK: Collins.

Sundberg, B., Nasholm, T. & Palmqvist, K. (2001). The effect of nitrogen on growth and key thallus components in the two tripartite lichens, *Nephroma arcticum* and *Peltigera aphthosa. Plant Cell Environment, 24,* 517–527.

Swindells, J. (2001). Thorow-wax as a pavement casual in Kensington. *BSBI News* 86, 50.

Tallent-Halsell, N. & Watt, M. (2009). The invasive *Buddleja davidii* (butterfly bush). *Botanical Review, 75,* 292–325.

Tansley, A. (1949). *The BritishIslands and their Vegetation.*Cambridge, UK: CambridgeUniversity Press.

Taylor, K. (1997). Biological flora of the British Isles: *Geum urbanum* L. *Journal of Ecology, 85,* 705–720.

Tesky, J. (1992). *Ceratodon purpureus.* USDA Forest Service Fire Effects Information System, http://www.fs.fed.us/database/feis/.

Thompson, K., Bakker, J. & Bekker, R. (1997). *The Soil Seed Banks of North WestEurope: Methodology, Density and Longevity.*Cambridge, UK: CambridgeUniversity Press.
Tiley, G. (2010). Biological flora of the British Isles: *Cirsium arvense* (L.) Scop. *Journal of Ecology, 98,* 938–983.

Tofts, R. (1999). Biological flora of the British Isles: *Cirsium eriophorum* (L.) Scop. (*Carduus eriophorus* L.; *Cnicus eriophora* (L.) Roth). *Journal of Ecology, 87*, 529–542.

Tremlett, M., Silvertown, J. & Tucker, C. (1984). An analysis of spatial and temporal variation in seedling survival of a monocarpic perennial, *Conium maculatum. Oikos, 43*, 41–45.

Tsuyuzaki, S. & Titus, J. (2010). Roadside vegetation in an oak forest, Oak Creek Wildlife Area, the Cascade Range, USA. *iForest, 3*, 52–55.

Tutin, T. (1957). Biological flora of the British Isles (*Allium ursinum* L.). *Journal of Ecology, 45*, 1003–1010.

UK National Ecosystem Assessment (2011).*The UK National Ecosystem Assessment: Synthesis of the Key Findings.* UNEP-WCMC, Cambridge.

Vick, C. & Bevan, R. (1976). Lichens and tar spot fungus (*Rhytisma acerinum*) as indicators of sulphur dioxide pollution on Merseyside. *Environmental Pollution, 11*, 203–216.

Wade, A. (1970). *The Flora of Monmouthshire.*Cardiff, Wales, UK: NationalMuseum of Wales.

Warren, J. (2009). Extra petals in the buttercup (*Ranunculus repens*) provide a quick method to estimate the age of meadows. *Annals of Botany,104,* 785-788.

Webster, M. & Brown, D. (1997). Preliminary observations on the growth of transplanted *Peltigera canina* under semi-natural conditions. *Lichenologist, 29*, 91–96.

Webster, J. & Weber, W.S. (2007). *Introduction to Fungi.* 3rd Edn. Cambridge, UK: CambridgeUniversity Press.

Wegner, C. & Hamburger, M. (2002). Occurrence of stable foam in the UpperRhineRiver caused by plant-derived surfactants. *Environmental Science and Technology, 36*, 3250–3256.

Wellnitz, T., Grief, K. & Sheldon, S. (1994). Response of macroinvertebrates to blooms of iron-depositing bacteria. *Hydrobiologia, 281*, 1–17.

Wells, T. (1975). The floristic composition of chalk grassland in Wiltshire. In L. Stern (Ed.), *Supplement to the Flora of Wiltshire* (pp. 99–125). Devizes, UK: Wiltshire Archaeological and Natural History Society.

Wells, T., Sheail, J., Ball, D. & Ward, L. (1976). Ecological studies on the PortonRanges: relationships between vegetation, soils and land-use history. *Journal of Ecology, 64*, 589–626.

Wilson, J. (1968). *The Control of Density in Some Woodland Plants.* PhD thesis, University of Lancaster, UK.

Winchester, V. (1984). A proposal for a new approach to lichenometry. *British Geomorphological Research Group Technical Bulletin, 33,* 3-20.

Winchester, V. (1988). An assessment of lichenometry as a method for dating recent stone movement in two circles in Cumbria and Oxfordshire. *Botanical Journal of the Linnean Society, 96,* 57-68.

Witkin, H., Moore, C., Goodenough, D. & Cox, P. (1977). Field-dependent and field-independent cognitive styles and their educational implications. *Review of Educational Research, 47*, 1–64.

Zohlen, A. & Tyler, G. (2004). Soluble inorganic tissue phosphorus and calcicole-calcifuge behaviour of plants. *Annals of Botany, 94*, 427–432.

OTHER USEFUL SOURCES

Glaves, P., Rotherham, I.D., Wright, B., Handley, C. & Birbeck, J. (2009).*A Report to the Woodland Trust. Field Surveys for Ancient Woodlands: Issues and Approaches.* Hallam Environmental Consultants Ltd., Biodiversity & Landscape History Research Institute, and Geography, Tourism & Environment Change Research Unit, Sheffield Hallam University, Sheffield.

Glaves, P., Rotherham, I.D., Wright, B., Handley, C. & Birbeck, J. (2009).*A Report to the Woodland Trust. A Survey of the Coverage, Use and Application of Ancient Woodland Indicator Lists in the UK.* Hallam Environmental Consultants Ltd., Biodiversity & Landscape History Research Institute, and Geography, Tourism & Environment Change Research Unit, Sheffield Hallam University, Sheffield.

Rotherham, I.D., Jones, M., Smith, L. & Handley, C. (Eds.) (2008). *The Woodland Heritage Manual: A Guide to Investigating Wooded Landscapes.* Wildtrack Publishing, Sheffield. ISBN 978-1-904098-07-2. 212pp

Rotherham, I.D. & Wight, B. (2011). Assessing woodland history and management using vascular plant indicators. *Aspects of Applied Biology, 108,* 105-112.

Rotherham, I.D. (2011).*A Landscape History Approach to the Assessment of Ancient Woodlands.* In: Wallace, E.B. (Ed.) *Woodlands: Ecology, Management and Conservation.* Nova Science Publishers Inc., USA, 161-184.

Rotherham, I.D., Glaves, P. & Wright, B. (undated) Evidencing Ancient Woodlands. *World of Trees, 20,* 42-44.

Vickers, A.D., Rotherham, I.D. & Rose, J.C. (2000). Vegetation succession and colonisation rates at the forest edge under different environmental conditions. *Aspects of Applied Biology, 58,* 351-356.

USEFUL WEBSITES

AUSTRALIAN NATIONAL BOTANIC GARDEN (WWW.ANBG.GOV.AU)

The Australian Government's AustralianNationalBotanic Garden maintains a web page on bryophyte ecology (www.anbg.gov.au/bryophyte/ecology-fire.html).

BOTANICAL SOCIETY OF THE BRITISH ISLES (WWW.BSBI.ORG.UK)

The BSBI maintains a list of axiophytes. These are defined by the Society as 'worthy plants' i.e. the 40% or so of species that arouse interest and praise from botanists when they are seen. They are indicators of habitat that is considered important for conservation, such as ancient woodlands, clear water and species-rich meadows. Lists of axiophytes by county are included on the website of the BSBI (http://www.bsbi.org.uk/html/axiophytes.html).

DEPARTMENT FOR FOOD, ENVIRONMENT AND RURAL AFFAIRS (DEFRA) (WWW.DEFRA.GOV.UK)

Defra maintains some useful summaries of wildlife legislation including the Weeds Act (1959) (www.defra.gov.uk/wildlife-pets/wildlife/management/weeds/index.htm).

INSTITUTE OF ECOLOGY AND ENVIRONMENTAL MANAGEMENT (WWW.IEEM.NET)

IEEM is an acrediting body for ecological professionals and maintains some online technical resources.

UNITED STATES DEPARTMENT OF AGRICULTURE (WWW.FS.FED.US)

The United States Department of Agriculture, Forest Service maintains an online Fire Effects Information System which provides useful information on the effects of fires on ecosystems (www.fs.fed.us/database/feis).

7. Appendices

Appendix 1: Ellenberg values

A detailed explanation of the Ellenberg scales and how they may be applied, and adapted, for use with the British flora is provided by Hill *et al.* (1999, 2004, 2007). Ellenberg defined seven major scales but two of these (continentality and temperature) are not considered satisfactory for use in the British climate (refer to Hill *et al.*, 2004). The remaining values are light (L), moisture (F from the German *Feuchtigkeit*), soil reaction (R), nitrogen (N) and salt tolerance (S). Bryophytes are provided with an additional category of heavy metal tolerance (HM) (Hill *et al.*, 2007).

The following points regarding the five scales are derived from Hill *et al.* (2004, 2007).

Light values

The full range of Ellenberg values for light is not represented in the British flora. For canopy trees, light values refer to the tolerance of the sapling stage of the life cycle. Hill *et al.* provide a scale of 1–9, where 1 (no British examples) represents 'deep shade' and 9 represents 'full sun' as in an open habitat such as a salt marsh. In the present guide, bird's nest orchid (*Neottia nidus-avis*) is an extreme example with an L value of 2, as encountered on the floor beneath a beech wood canopy in summer. For bryophytes the light scale is extended to include 0 (plants which live in darkness e.g. ghostwort (*Cryptothallus mirabilis*)).

Moisture values

Response to moisture is presented on a scale of 1–12, where 1 is assigned to plants found on soils of extreme dryness and 12 refers to submerged species such as greater water-moss (*Fontinalis antipyretica*). An indicator of extreme dryness is the white rock-rose (*Helianthemum appeninum*) which grows on steep, limestone grasslands such as Brean Down in Somerset.

Soil reaction values

On a scale of 1–9,R refers to environmental acidity (of the soil or aquatic medium) which would ordinarily be measured by pH. It is important to note however that R values are not the same as pH. They are an arbitrary scale reflecting soil pH though not directly based on measurements (refer to Hill *et al.*, 1999). On the R scale, 1 indicates extreme acidity (plants never found on weakly acid or basic soils) and 9 is an indicator of basic reaction (plants always found on calcareous or other high pH soils).

Nitrogen values

On a scale of 1–9, these are a general indication of a plant's preference for soil fertility. Low N values correspond to plants with a high stress tolerance (*sensu* Grime *et al.*, 1997) and high values with a low stress tolerance (refer to Grime *et al.*, 2007 for a discussion of the stress tolerator strategy).An N value of 1 indicates plants characteristic of extremely infertile sites (e.g. bog plants such as sundew *Drosera rotundifolia*) while an N value of 9 refers to plants which may be regarded as indicators of extremely nutrient-rich situations such as cattle resting places (e.g. ruderal species such as broad-leaved dock, *Rumex obtusifolius*). The N scale for bryophytes terminates at 7, as 8 and 9 correspond to conditions where bryophytes are crowded out by vascular plants. The 'bonfire-moss' (*Funaria hygrometrica*) is found in richly fertile places at the upper end of the scale for bryophyte survival (N value 7).

Salt tolerance values

These range from 0 to 9 and are based on a plant's tolerance to salt. Species scoring 0 are absent from saline sites whereas those scoring 9 occur on sites where seawater evaporates to leave extremely saline conditions such as may be found at the lower edge of a salt marsh. Bryophytes are not accorded salt tolerance values greater than 5 (which includes obligate halophytes).

HEAVYMETAL TOLERANCE

This scale is used for bryophytes only and extends from 0 to 5, where 0 includes species which are absent from substrates with moderate or high concentrations of heavy metals and 5 includes species which, in the British Isles, are confined to substrates with a moderate or high metal content.

Appendix 2: Photographs to Test Your Abilities as a Plant Detective

Take a while to study the test photograph then deduce as much as you can about the habitats portrayed. Suggested answers are on the last page. The cautionary photograph is included as a warning to use botanical indicators with care.

Test Photograph

Cautionary Photograph

Answers to test

Test Photograph

This photograph shows a Cotswold landscape, part of the Snowshill valley (Worcestershire), with fields, small woodlands and the 12[th]century St Eadburgha's church. Hawthorn (*Crataegus monogyna*) is in flower at various points, which indicates that the photograph was taken in May or early June. Blackthorn (*Prunus spinosa*), which also has white flowers, would not be showing leaves whilst flowering and fewer trees would be in leaf when blackthorn is in flower in April. The horizon is marked by a line of mature, broad-leaved woodland and Broadway Tower (a folly). The nature of the woodland is difficult to deduce from the picture alone but large trees are clearly present and the presence of flowering hawthorn at the edge indicates a developed shrub layer. Clearly this is not a plantation and therefore it may support a woodland herb flora too. It can be readily deduced that this woodland merits further investigation. In fact it supports quite a diverse woodland herb flora.

A line of hawthorn marks a hedge leading off at 45°from the central woodland strip. The difference between these two lines of trees/shrubs is notable. The central line is tall, relatively thick and varies in colour and texture; it sits in a shallow valley. The smaller 45°boundary is relatively thin by comparison and uniform in colour (most of the shrubs have white flowers). It is reasonable to conclude that the smaller hedge is the younger line, consisting of a row of planted hawthorns of relatively uniform age (i.e. a newer boundary than the central line). The central line itself shows no white flowers and includes much larger trees of varying shapes and forms i.e. it is a species-diverse feature of greater age.

In fact, the central line consists of two strips of shrubs with mature and veteran trees and marks the line of a mediaeval pack-horse route, Coneygree Lane, which climbed the ridge at this point, leaving the valley by St Eadburgha's church. The strips of woodland conceal a stream and contain ancient woodland indicator species (wood melick *Melica uniflora*, ramsons *Allium ursinum*, bluebell *Hyacinthoides non-scripta*, herb paris *Paris quadrifolia*, early purple orchid *Orchis mascula*, pendulous sedge *Carex pendula*, wood sedge *Carex sylvatica*, goldilocks *Ranunculus auricomus*, wood anemone *Anemone nemorosa*, sanicle *Sanicula europaea*). Though the fine detail is not visible from the picture, the difference in character of these two boundaries is clear to see even at a distance and the wary plant detective would immediately consider investigating the lane further.

The field in the distance shows yellow patches of flowering buttercups (*Ranunculus* species), also confirming the time of the photograph to be May/June. This field has an irregular surface. This may indicate some form of former, small-scale quarrying activity. Such ground disturbance often leaves areas of thinner mineral soils that may support less-common flora. The variation in texture and colour in this field argues against an agriculturally improved grassland sward and in favour of a species-rich, unimproved grassland flora. If a choice had to be made from this photograph, the buttercup field would be the choice for more detailed grassland investigations. The field boundary in the centre of the picture, one-third down from the horizon shows many white streaks. These are protective tubes for young saplings. Some effort has therefore been made to extend the woodland out from the central strip, indicating that conservation planning has been applied to the landscape. Someone is interested in its ecological quality.

This is a landscape with some large trees on its field boundaries. Some of these might be veterans and are worth further investigation. The large tree at the left hand edge of the planting area is in fact a beech (*Fagus sylvatica*) with the date 1876 carved on its trunk.

The church and cottages visible in the mid-ground of the photograph appear to be constructed of natural Cotswold stone and would therefore be good places to look for lichens. The churchyard clearly contains gravestones which also provide suitable habitats for such species. Churchyards often support remnant ancient grassland and veteran trees (e.g. yew *Taxus baccata*) and are worth investigating for such features. Also in the mid-ground is a sports field. In this area, the uniform texture of the sward hints at a mowing regime. In an area surrounded by pastures and woodland, mowing might encourage certain elements of flora, less common in the surrounding landscape.

The field in the foreground is crossed by two tyre tracks. The tracks pass between two small patches of nettle (*Urtica dioica*) and there is also evidence of thistles in the field. Otherwise the sward appears relatively uniform in colour and texture. These clues indicate an improved grassland sward with accrued phosphate-rich patches. The reason for these local patches is unclear from the photograph. Perhaps they indicate places where animal faeces accumulated during supplemental winterfeeding, but in the absence of cropping to remove them such features may last for centuries and could be difficult to explain without further information.

Cautionary Photograph

This photograph is included as a warning to take care with botanical indicators. The evidence is as follows:

A road verge with a crash barrier, road signs, street lamps (which need cables), a cracked black-top footway, a bank with some trees at the bottom.

The conspicuous indicators are:

Creeping thistle (*Cirsium arvense*):	Indicates fertile soil
White clover (*Trifolium repens*):	Indicates nutrient enrichment
Oat grass (*Arrhenatherum elatius*):	Indicates absence of grazing, infrequent cutting.

The evidence adds up to an average botanically impoverished road verge but in the foreground, just beyond the wooden fence support, is a lizard orchid (*Himantoglossum hircinum*), one of a significant population on this particular site near Bristol. This is apparently a constructed, highly disturbed verge but some unknown factor renders it suitable for lizard orchid. The conclusion is that the niche requirements of lizard orchid are not yet clear.

Acknowledgements

Carol E. Jenner, David Gledhill, Chris Bentley and B. K. Edward are acknowledged for their helpful comments on early versions of this manuscript.
Thanks to Rosemary and members of the Cheltenham U3A for their encouragement.

www.ingramcontent.com/pod-product-compliance
Ingram Content Group UK Ltd.
Pitfield, Milton Keynes, MK11 3LW, UK
UKHW050614260726
13967UKWH00008B/2862

9 781904 098362